VER

Ce

ist

y

SECONDS

19

27

9

99

Concrete Detail Design

The Concrete Society

The Architectural Press: London

First published in 1986 by the Architectural Press Ltd,
9 Queen Anne's Gate, London SW1H 9BY

BRITISH LIBRARY CATALOGUING IN PUBLICATION DATA

Concrete detail design. – (Architectural
    Press library of design and detailing)
    1. Reinforced concrete
    I. Concrete Society
    624.1'8341   TA444

ISBN 0-85139–795–6

Typeset by Crawley Composition Limited, Stephenson Way,
Three Bridges, Crawley, Sussex RH10 1TN
Printed and bound in Great Britain by
Biddles Ltd, Guildford and King's Lynn

This book was produced by a working party of the Concrete Society's Structural Committee, under the chairmanship of John Mason. The members were:

JOHN MASON M.A., C.Eng., M.I.Struct.E. graduated in 1970 from Jesus College, Cambridge, and gained his initial training with F. J. Samuely & Partners, then Kenchington Little and Partners. This was followed by a spell as a research manager at C.I.R.I.A. then Resident Engineer and Senior Engineer for S. B. Tietz & Partners. Now a partner of Alan Baxter & Associates, he has a wide experience of building design in concrete and all other materials, taking a particular interest in the interaction between architects and structural engineers.

JOHN DUELL B.A., B.Arch., R.I.B.A. graduated in 1964 from the University of Illinois, U.S.A. He then came to the U.K. and worked in the G.L.C. Architect's Department. In 1965 Mr Duell won a major architectural competition and formed his own practice. In 1979 he became a partner in Hurley Porte & Duell. He has written extensively on materials, products and construction for *The Architects' Journal*.

DAVID PRICHARD C.Eng., F.I.C.E., M.Cons.E. began his professional career with Sir Alexander Gibb & Partners and moved to Richard Costain before joining Harris & Sutherland in 1965. He worked initially on design related to universities at Essex and Bath and was appointed an associate partner in 1970, responsible for a wide range of structural projects. He became a partner in 1978, and continues to be involved in structural projects using a variety of materials.

RON SLADE B.Sc., C. Eng., M.I.Struct.E. joined Kenchington Little & Partners in 1963 and graduated from The City University in 1967. After initial design office experience he was Resident Engineer on a number of building and civil engineering sites. In 1972 he was appointed Executive Engineer with Kenchington Little & Partners and has since been responsible for major refurbishment projects and for the structural design of a number of large buildings.

DEREK WINSOR B.Sc., C.Eng., M.I.C.E., M.I.Struct.E. graduated in 1976 from Loughborough University of Technology from where he joined Mott, Hay & Anderson, Structural and Industrial Consultants and has since gained wide experience in the design and construction supervision of a varied range of both commercial and industrial projects in the U.K. and overseas.

Founded in 1966, the Concrete Society brings together all those with an interest in concrete – to promote excellence in its use, to encourage innovation and to exchange knowledge and experience across all disciplines for the public good.

Reinforced concrete is used in some form or other in almost every building, yet repair and maintenance problems with reinforced concrete construction are increasing. Problems such as spalling and cracking concrete, reinforcement corrosion, excessive deflection, cracking finishes and partitions and inadequate support of claddings derive principally from a lack of clear understanding by designers – both architects and structural engineers – of the criteria to be met in producing durable and robust building designs.

It is where the design responsibilities of the two disciplines meet that problems arise – first with construction on site, and later with durability. All too often these aspects of construction are not explored fully at the design stage, but left to the contractor, who has to sort them out on site.

The Concrete Society, through a working party of its Structural Committee, identified a lack of published guidance on realistic minimum sizes and sensible profiles for in situ concrete to enable architects and engineers to achieve designs that will perform well. Rather than spending years producing a comprehensive manual covering all aspects and forms of concrete construction, it was felt more important to produce immediate guidance on recurrent problems. Therefore the scope of this book has been deliberately restricted to common difficulties in cast in situ reinforced concrete-framed construction.

Other aspects of in situ framed construction, such as beams, stairs and lift shafts, do not usually cause so many problems and are only referred to in passing.

The guidance given here assumes the use of normal materials and construction techniques. Where circumstances dictate, it may be possible to use smaller sizes than those recommended by employing special materials, for example stainless steel reinforcement.

Simplicity

This first chapter considers broad constraints on designing with in situ concrete.

Apparently simple details can produce disproportionately complex junctions, which are often overlooked during design. These junctions must be identified at an early stage and investigated fully to produce functional and practical solutions.

This problem occurs most frequently at perimeter beam-column intersections where the elevations have been designed to be visually interesting. An example will illustrate the point: the elevation at small scale looks fairly straightforward (Figure 1.1) and the solution to various details has been found – but only in isolation (Figures 1.2–1.4). The result is a complex intersection of reinforced concrete beams, columns and walls (Figure 1.5). Not only is the structural design difficult, but brick cladding details and dpcs are complicated. This is exacerbated by having to design for relative movement between the concrete frame and the brick cladding.

Figure 1.1   Elevation of junction where the vehicle ramp to the rooftop car park meets the plant room.

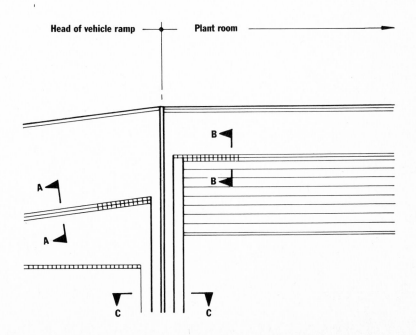

On site this creates a problem for the shuttering carpenter, steel-fixer and bricklayer. Without proper investigation by the design team there are the dangers of inadequate measurement at tender stage and cost problems.

Early liaison between the architect and the engineer is essential in order to identify such junctions and to decide whether to meet the design demands of the problem or to avoid it. All too often the need for simplicity is not recognised when designing typical sections, especially when there is a rush to get a project on site.

Figure 1.2   Modelling by corbelling brickwork, section A-A.

Figure 1.3   Corbelling above the ventilation grille, section B-B.

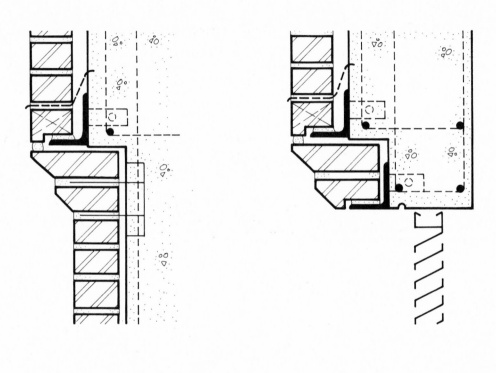

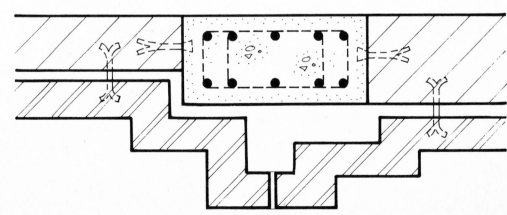

Figure 1.4   Complex plan resulting from set-backs above, plan C-C.

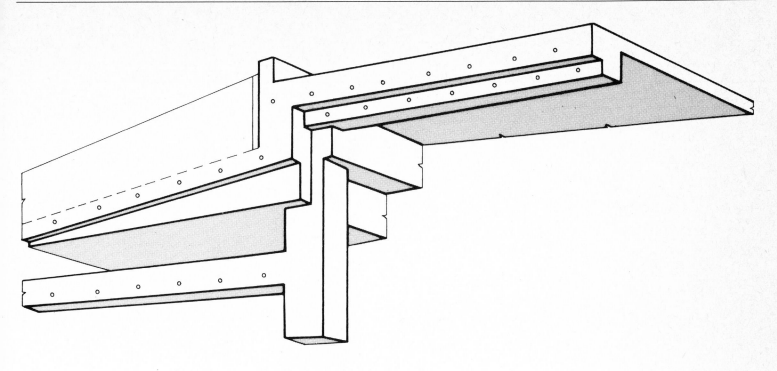

Figure 1.5   Complex in situ framing.

Accuracy, Tolerances and Fit

Chapter 2 looks at problems of setting out and achieving easily buildable designs.

Accuracy, tolerances and fit in buildings have been recognised as critical, but there is no generally accepted guidance for designers. BS 5606 *Code of Practice for Accuracy in Building* advises on the probable accuracy in some 'normal' forms of construction under 'normal' conditions. The code's statistical approach is developed further and illustrated in C.I.R.I.A.'s Technical Note 113. For general aspects of accuracy see the series of articles in *The Architects' Journal* 'Design for Better Assembly' (numbers 25.7.84, 1, 8 & 15.8.84 and 5.9.84).

In situ reinforced concrete can be one of the least precise structural materials and it is important to allow for this when designing in situ concrete-framed buildings.

2.1   Setting Out

Plan tolerances are usually specified and deviations measured from a building grid, which has its own inaccuracies (Figure 2.1). Vertical setting out is similar,

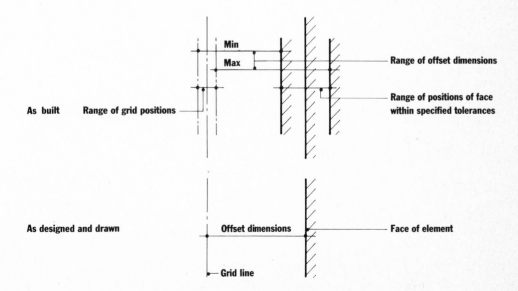

Figure 2.1   Setting out relative to the grid – as designed and built. Grid and face position inaccuracies can combine to make offset dimensions variable.

specifying levels relative to a datum point (usually an OS benchmark). Although the grid or datum level system is convenient for dimensioning building elements on drawings and their positioning on site, it can exacerbate problems of tolerance and fit. This is because the grid system defines an *absolute* position for each element, whereas in fact the *relative* position of elements is often critical (Figure 2.2). The maximum permitted tolerance for the elements may not be compatible with achieving good fit.

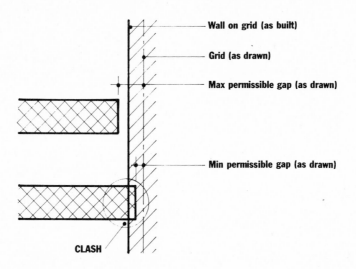

Figure 2.2   Grid inaccuracies lead to a clash of components.

The approach normally adopted on site for in situ concrete-framed buildings is a compromise between achieving theoretical offsets from the grid and establishing a practical and buildable relationship between the frame as built and the rest of the fabric – the cladding, windows, partitions, services and so on. For example, if the perimeter of the frame is constructed out of plumb, the cladding will be adjusted to achieve the best possible line rather than maintaining an offset from the grid (Figures 2.3, 2.4).

Figure 2.3   Stage one – the frame is built relative to the grid and levels.

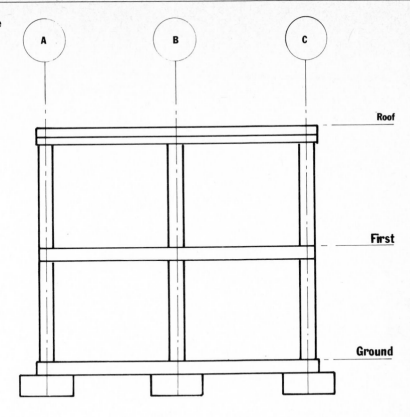

Figure 2.4   Stage two – the rest of the fabric tends to be built relative to the frame. The grid is ignored.

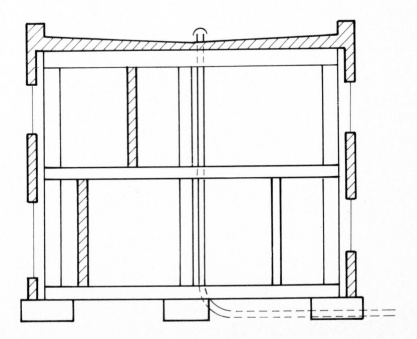

2.2   Assembly of Components and Materials

The various components and materials used in construction differ widely in how easily they may be adjusted and fitted to their neighbours. There are three basic groups: prefabricated, cut-to-fit and in situ moulded components.

2.2.1   Prefabricated Components

These include windows and other joinery, steelwork, metalwork, services fittings and plant, precast concrete and the faces of brickwork and blockwork.

2.2.2   Cut-to-fit Components

Among these components are the length and height of brickwork and blockwork, carpentry, tiling, felting and plasterboard.

2.2.3   In Situ Moulded Components

These include in situ concrete, asphalt and wet plastering.

If one or both components are cut to fit, junctions will normally be free from problems of fit or tolerance. Note that 'traditional' construction generally uses this type of junction (Figure 2.5).

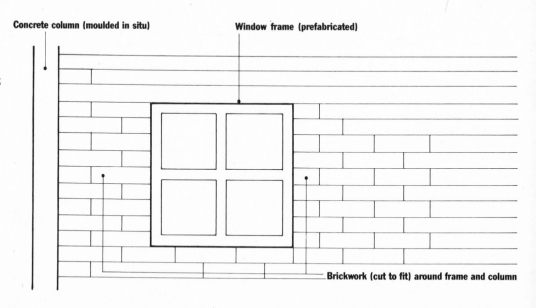

Figure 2.5   Functions of cut-to-fit components. Ideally openings in masonry should suit the brick or block module.

Concrete column (moulded in situ)   Window frame (prefabricated)

Brickwork (cut to fit) around frame and column

Junctions between various prefabricated components, and between in situ moulded and prefabricated components when the in situ moulded component is built first, need to take account of accuracy and fit (Figure 2.6). Here the connections between elements should be designed to allow sufficient tolerance or the design should be modified to avoid such junctions. In interfaces between in situ moulded and prefabricated components, measurements on site of the in situ elements before prefabrication may alleviate problems, provided the longer time and higher cost needed for this process are acceptable (Figure 2.7).

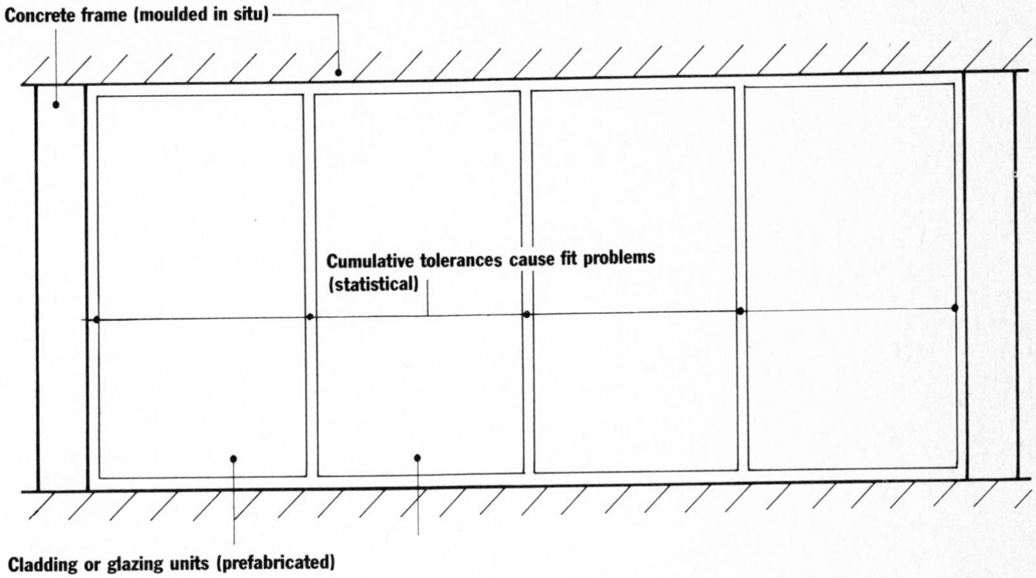

Concrete frame (moulded in situ)

Figure 2.6   Problems of cumulative tolerance where prefabricated components are built after in situ moulded components.

Cumulative tolerances cause fit problems (statistical)

Cladding or glazing units (prefabricated)

Figure 2.7   Prefabrication started after in situ moulding is complete to fit components to openings as built.

Concrete frame (moulded in situ)

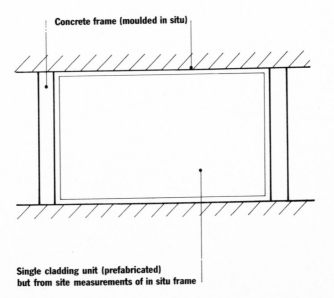

Single cladding unit (prefabricated) but from site measurements of in situ frame

2.3   Common Problems with In Situ Frames

Where the concrete frame abuts external cladding, 'normal' permissible deviations to BS 5606 will allow too much variation along the edge of the frame for straightforward construction of plumb, flat brick or block cladding (Figure 2.8). Either the contractor must be made aware of the need to achieve accurate construction or the frame-cladding junction should be detailed to eliminate the variations (at extra cost). This problem is acute where the cladding is carried on an angle support (Figure 2.9). This is discussed in more detail in Chapter 9.

Figure 2.8   The 'normal' deviation of the edge can create problems in the plumb construction of infill masonry.

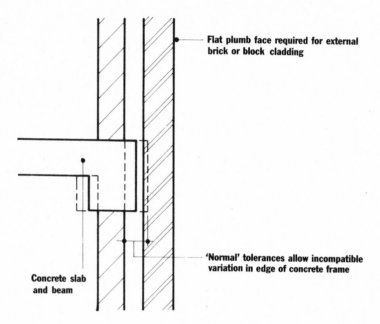

Flat plumb face required for external brick or block cladding

'Normal' tolerances allow incompatible variation in edge of concrete frame

Concrete slab and beam

Permissible deviations to BS 5606 of '±x' are incompatible with deviations to BS 5655: Part 5 for lift shafts of +25 to 0 mm. Here the *structural* dimensions of the shaft should be increased by 25 mm and 'special' tolerances in plan of ±12 mm specified (Figure 2.10). Extreme care is needed to avoid confusing lift specialists' and structural drawings, which will have different dimensions.

'Special' requirements, that is outside BS 5606 normal tolerances, also need to be considered for floated floor slabs for warehousing and to ensure adequate structural falls where required on roof slabs. For example, the minimum fall may be specified as 1:40 so the minimum fall as built is between 1:60 and 1:80.

Figure 2.9   Where the frame edge is out of line with the floors below, vertical masonry may not meet the supporting angle accurately enough and may need cutting, left, or lack support, right.

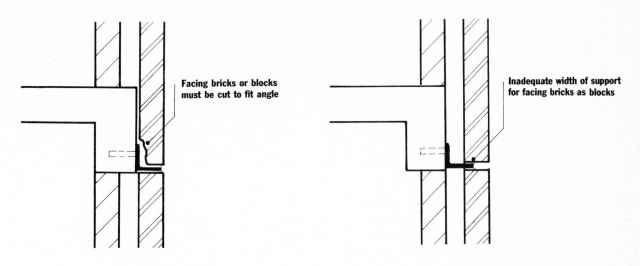

Facing bricks or blocks
must be cut to fit angle

Inadequate width of support
for facing bricks as blocks

Figure 2.10   Shaft dimensions, left, as specified by the lift specialist and, right, as shown on structural drawings.

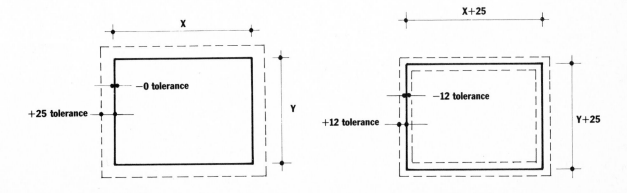

X

−0 tolerance

+25 tolerance

Y

X+25

−12 tolerance

+12 tolerance

Y+25

2.4   Design Checklist

As the design of a concrete-framed building is developed, the following questions should always be borne in mind:

o   is the grid and datum level setting out system likely to lead to a clash between elements and, if so, can the clash be designed out or should 'special' tolerances be specified?

o   if the worst combination of deviations occurs, does it matter?

o   is any junction between components likely to cause tolerance problems and, if so, can one or both of the components be cut to fit, or should a statistical assessment of tolerances and fit be made and designed for?

o   have all unusual aspects been identified?

Movement Joints

Chapter 3 reviews causes and effects of structural movement and considers methods of dealing with them.

In building design distress can be caused by building movement or by the relative movement of components. Movements in building frames can be classified as follows:

- vertical movement caused by differential settlement (Figure 3.1)
- horizontal movement due to temperature and shrinkage (Figure 3.2)
- vertical movement caused by load, shrinkage and creep (Figure 3.3)

In addition, relative movement between the structure and infill components should be considered, notably expansion in clay brickwork and shrinkage in blockwork and calcium silicate brickwork (Figure 3.4).

Vertical movement due to differential settlement or horizontal movement caused by temperature and shrinkage may mean that full structural joints are needed to divide the structure into separate sections. Vertical movements due to load, shrinkage and creep, and relative movement, lead to localised joints. These

Figure 3.1 Vertical movement due to differential settlement.

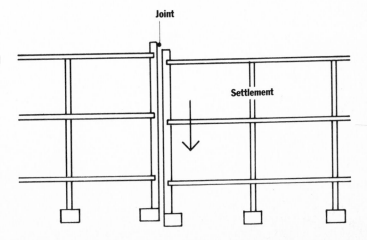

Figure 3.2   Horizontal thermal and shrinkage movement.

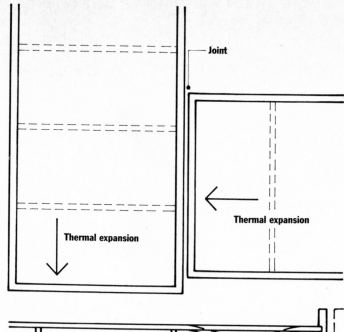

Figure 3.3   Vertical movement due to load, shrinkage and creep.

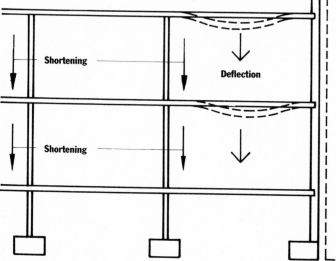

Figure 3.4   Expansion in clay brickwork.

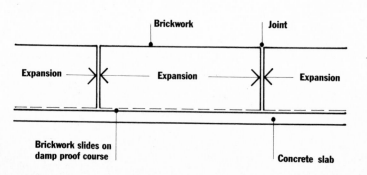

movements, where they involve masonry and reinforced concrete frames, are also discussed in Chapter 9.

3.1 Sources of Movement in Concrete

These sources are listed below in order of importance.

3.1.1 Long-term Shrinkage

Reinforced concrete shrinks with loss of moisture and internal chemical changes. The effect is most pronounced early in the structure's life but, depending on the thickness of the section, the percentage of reinforcement and the relative humidity, the total shrinkage may still be only 70–90 per cent complete after three years. It may cause an element to shorten or to bow and twist also.

3.1.2 Temperature and Solar Radiation Effects

Movement is caused by changes in the seasonal and daily outdoor temperature, by variations in indoor temperatures and by the effect of solar radiation on exposed surfaces. The most important factor to note when estimating the probable amount of movement is whether the structure is insulated. Thermal inertia of individual elements should also be considered.

3.1.3 Elastic Deformation and Creep due to Load

The initial application of load causes immediate deflection. Under sustained loading the deflection will increase due to the gradual deformation of the materials. This phenomenon is known as 'creep'. Vertical elements are also subject to creep, causing shortening.

3.1.4 Early Thermal Movement

High temperatures are generated by the hydration of cement as concrete sets. This causes an initial expansion, and in certain circumstances the subsequent cooling and shrinkage may cause the immature concrete to crack.

3.2 Design for Movement

There are no hard and fast rules for minimising the effect of movement and no code of practice for specific guidance. However, it is important that the design team should consider the effects of the expected movements early in design so that provision for movement can be properly incorporated in the structure, cladding and finishes. The position of movement joints will be a matter of judgement but the principles illustrated below generally apply. The consensus view is that movement joints often introduce significant problems of inspection and maintenance, so the number of joints should be kept to a minimum.

3.2.1   Building Shape

In some buildings it is clear that the effect of movement will be concentrated at certain positions where there are abrupt changes of shape or direction, and full structural joints must be incorporated (Figure 3.5).

Figure 3.5   Joint at 'natural' break in building.

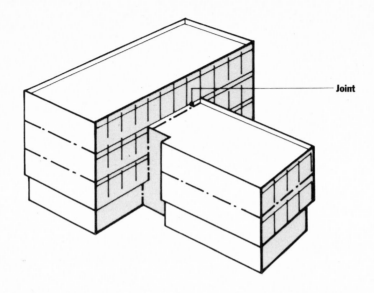

3.2.2   Joint Spacing

The spacing of joints, and indeed the decision on whether any joint is required at all, depends on many factors, but it is not unusual to have joints at 50–70 m centres in fully-framed construction (Figure 3.6). Bearing in mind that concrete

Figure 3.6   Spacing of major movement joints, located to avoid changes in structure, such as set-backs.

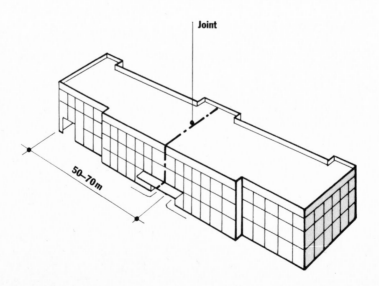

made from modern cement shrinks more, many engineers would argue that joints should be more closely spaced. If the construction is more rigid, joints should be at closer centres (Figure 3.7).

Figure 3.7   The stiffer the structure, the closer the movement joints.

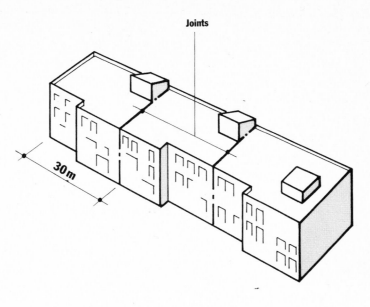

3.2.3   Movement Joints in Top Decks and Roofs

If a building has an uninsulated upper deck exposed to solar radiation, it may be advisable to have joints at closer centres than on the rest of the structure (Figure 3.8).

Figure 3.8   Joints may need to be closer where the effects of the movement noted earlier, particularly thermal movement, are greater.

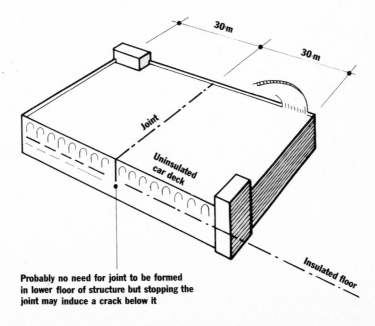

Where a roof, even one that is insulated, is supported on closely spaced masonry walls, the joints may need to be spaced at 15–20 m centres (Figure 3.9).

Figure 3.9 A concrete roof on brick walls requires relatively close joint spacing (based on *Civil Engineering Reference Book,* edited by L. S. Blake, Newnes-Butterworth).

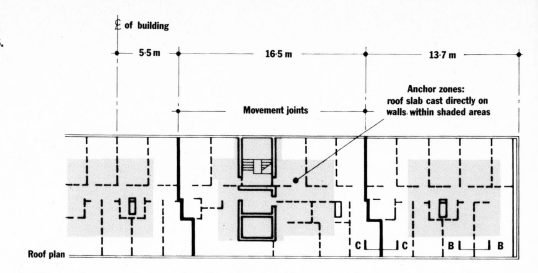

Roof plan

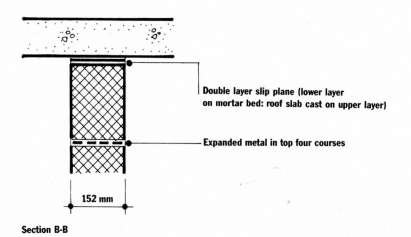

Section B-B

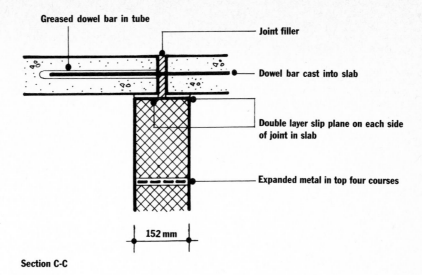

Greased dowel bar in tube

Joint filler

Dowel bar cast into slab

Double layer slip plane on each side
of joint in slab

Expanded metal in top four courses

152 mm

Section C-C

3.2.4 Foundation Movement and Settlement

Movement can result from the weight of the building applied to the ground. The effect can be more pronounced if variations in the loading cause differential settlement. The ground itself may cause movement due to changes in ground water conditions or mining subsidence.

As an alternative to providing vertical separation of parts of the structure, pairs of joints can be formed to produce an articulated joint to accommodate differential settlement (Figure 3.10).

Figure 3.10   Articulated joints in structures which load the ground unevenly (based on *Design for Movement in Buildings* by S. J. Alexander and R. M. Lawson, C.I.R.I.A., 1981).

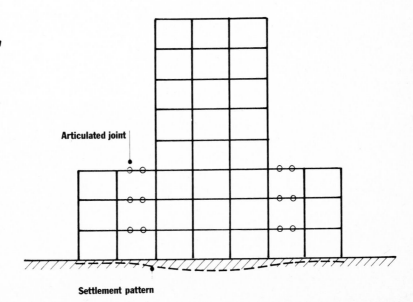

Articulated joint

Settlement pattern

3.2.5 Movement in Long Buildings with Two Cores

Long buildings with two cores (Figure 3.11) present a dilemma. Structural stability is best achieved by using the floor slabs to span horizontally to transfer horizontal loads to the cores. However, the cores will tend to restrain longitudinal movements (especially contraction) of the floor slabs. Adequate reinforcement must be provided to control the subsequent shrinkage cracking, but not so much as to induce unacceptably large shear forces in the lowest storeys of the cores.

The alternative is to introduce a movement joint, but this breaks the structural continuity (Figure 3.12). Both ends of the building are then designed to be structurally independent, with each core normally needing to act in torsion to provide stability.

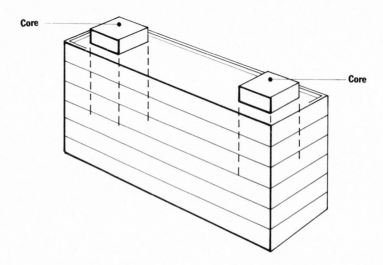

Figure 3.11   For a building having two cores it may be possible to avoid movement joints.

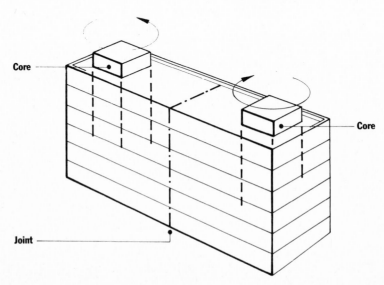

Figure 3.12   Movement joint leaving two virtually independent structures. Cores must be designed for additional torsional loads.

3.3   Joint Detailing for Horizontal Movement

Two basic arrangements can be used. These are the halved joint (Figure 3.13) and the split column (Figure 3.14). Detailing varies with the surfaces involved. When possible at roof level, use an upstand detail to prevent ingress of water (Figure 3.15). Beware joint-filler materials which are not compressible enough.

Flush joints in roof decks such as those required in car parks are much more difficult to deal with. The simplest way is to use a mastic filler material but often the filler cannot accommodate the expected movement, and this approach is not recommended (Figure 3.16). One option is to provide a gutter under an open joint. A proprietary system may often be the best solution (Figure 3.17). Its initial cost will be high but in the long run it may be money well spent. Flush joints in internal floors are fairly simple to design (Figure 3.18).

Figure 3.13   Halved joint, which needs very careful detailing and construction.

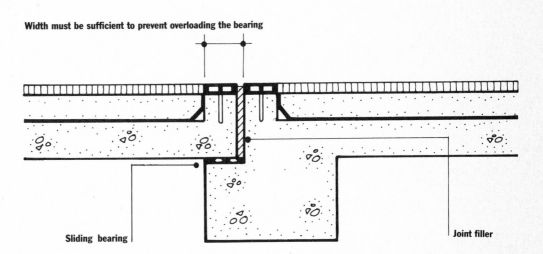

Figure 3.14   Split columns are fairly foolproof, so they are preferable to a halved joint.

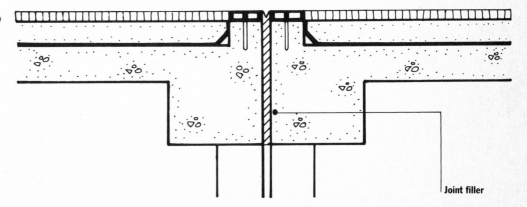

**Figure 3.15   Capped upstands at a movement
joint on the roof, the preferred roof detail.**

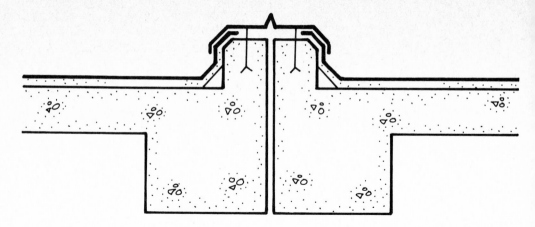

**Figure 3.16   Fillers to flush joints tend to fail
because they cannot accommodate enough
movement.**

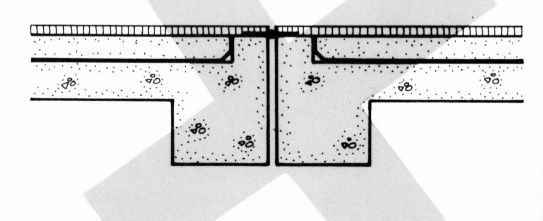

Figure 3.17   Proprietary flush joint: expensive but preferable where flush joints cannot be avoided.

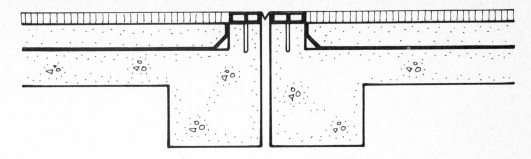

Figure 3.18   Without weatherproofing problems, flush joints are straightforward inside.

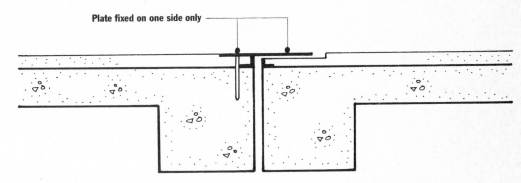

Plate fixed on one side only

3.4   Detailing Cladding and Partitions for Movement

It is very important that movement joints should pass right through the structure, cladding and finishes (Figure 3.19). Cladding and partitions should be detailed to accommodate frame-shortening as described above, and to allow for the deflection

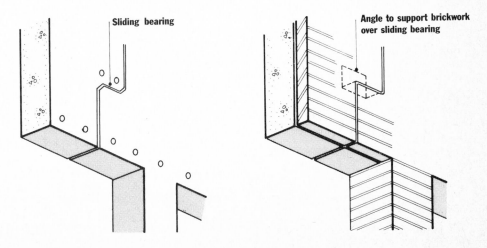

Sliding bearing

Angle to support brickwork over sliding bearing

Figure 3.19   Example of a joint in a beam showing how the joint must also pass through cladding, itself with extra supports as required.

of supporting slabs or beams (Figure 3.20). In two-way flat slabs deflections are greater than in one-way slabs (Figure 3.21).

Cladding and partitions supported by a flexible structure should have joints at fairly close spacing. In particular flat slabs or long spans need careful attention, and require close liaison between architect and engineer.

Figure 3.20   Stiff partition built without joints on a flexible structure.

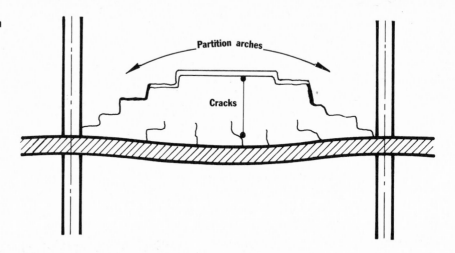

Figure 3.21   Greater deflection occurs in two-way slabs.

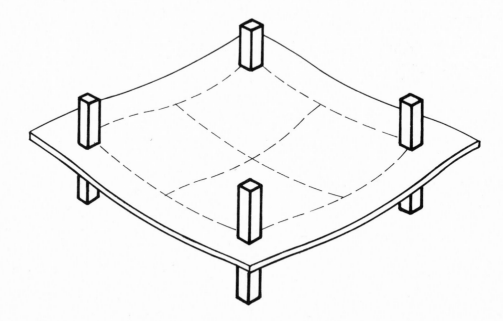

Selection of Floor Construction

Chapter 4 covers types of construction of floors and outline sizing for the three basic types – one-way, two-way and flat slabs.

These notes comparing structural and functional properties of common floor constructions cannot provide a complete prescription for floor design, and can only give guidelines on likely floor depths for various spans. Each job has to be considered on its merits. However, the criteria provided give a general guide to choosing floor construction and an agenda for discussion with the engineer.

4.1   Structural Arrangement of Slabs

There are three general arrangements:

One-way spanning slabs
These span between lines of supporting beams and columns or walls, which are usually parallel (Figure 4.1).

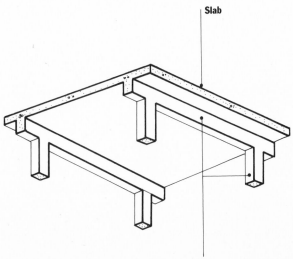

Slab

Stiff supporting beams and columns

Figure 4.1   One-way slab.

Two-way spanning slabs

These are supported on a rectangular grid of beams, with columns at beam intersections – the system is not suited to irregular column layouts (Figure 4.2).

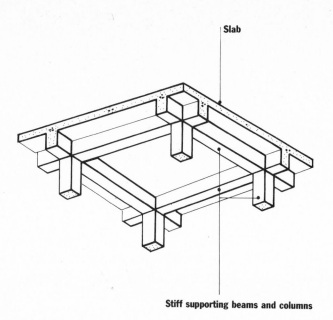

Slab

Stiff supporting beams and columns

Figure 4.2   Two-way slab.

Flat slabs

These are carried directly on columns with no beams (Figure 4.3). This system can accommodate irregular column layouts more easily. Perimeter columns often need

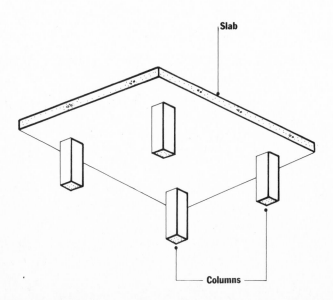

Slab

Columns

Figure 4.3   Flat slab.

to be more closely spaced than internal ones to provide enough stiffness to support the cladding. Alternatively, upstand or downstand beams may be used but these complicate and slow down construction.

4.2 Performance

Comparisons between the performance of floor constructions are shown below for variations of the three basic types. For each construction the following properties are rated between 'poor' and 'good':

o   fire resistance of the floor
o   sound attenuation of the floor and resistance to flanking transmission between rooms
o   ease of accommodating small services openings – for example for individual pipes or conduits – up to about 250 mm
o   ease of accommodating large service openings – for example for ventilation ducts or groups of pipes – up to about 1 m
o   ease of accommodating major openings – for example for lift shafts, stairwells or principal service ducts
o   ease of accommodating horizontal services distribution close to the slab soffit
o   ease of making fixings to the slab soffit for suspended ceilings, services, etc
o   ease of fixing heads of partitions to the slab soffit
o   ease and speed of constructing the slab
o   ease with which perimeter cladding support details may be accommodated.

Comparisons have not been made between costs, construction (self) weights, imposed load capacities or influences on storey heights. It is for the designers of each scheme to assess the importance of these factors in each case. (The merits and drawbacks of ribbed floor construction are discussed in Chapter 5 and the problems of forming holes through floor slabs in Chapter 6.)

4.3 Performance: One-way Spanning Slabs

Solid slab
The downstand beam does not have to be the same width as the column (Figure 4.4).

Hollow pot slab
The horizontal service distribution, and construction ease and speed can be improved if the downstand beam is replaced by a wide shallow beam to give a level soffit (as with the ribbed slab, see below) (Figure 4.5).

Figure 4.4   (left) One-way slab – solid; (right) performance rating.

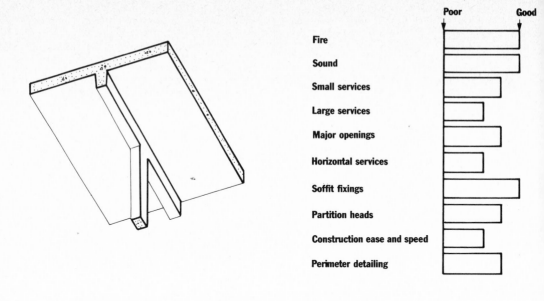

Figure 4.5   (left) One-way slab – hollow pot; (right) performance rating.

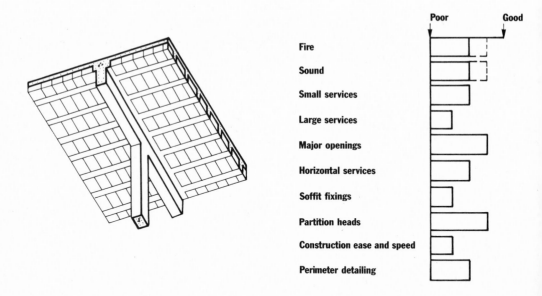

Ribbed (trough) slab
The horizontal service distribution, and construction ease and speed will be hindered if downstand beams are introduced. This is not suited to a square grid of columns (Figure 4.6).

Figure 4.6   (left) One-way slab – ribbed/trough; (right) performance rating.

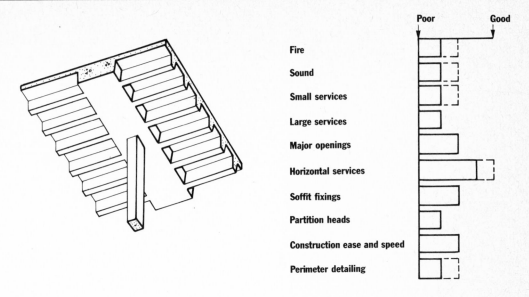

4.4   Performance: Two-way Spanning Slabs

Solid slab

Downstand beams do not have to be the same width as the columns (Figure 4.7).

Figure 4.7   (left) Two-way slab – solid; (right) performance rating.

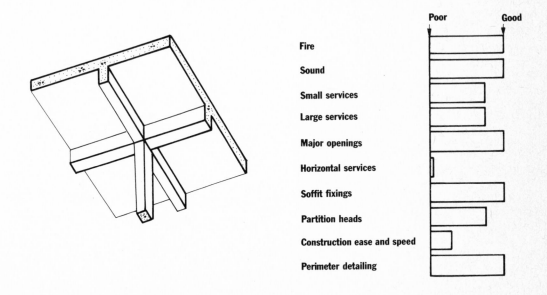

Ribbed slab using lightweight block formers
The horizontal service distribution, and construction ease and speed can be improved if the downstand beam is replaced by a wide shallow beam to give a level soffit. In this case assess the depth as for a coffered flat slab (Figure 4.8).

Figure 4.8 (left) Two-way slab – ribbed with block formers; (right) performance rating.

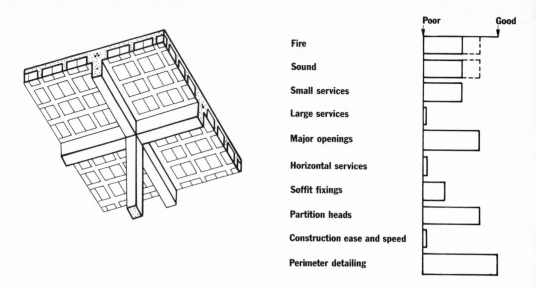

	Poor	Good
Fire		
Sound		
Small services		
Large services		
Major openings		
Horizontal services		
Soffit fixings		
Partition heads		
Construction ease and speed		
Perimeter detailing		

Coffered (waffle) slab
Assess the depth as for a coffered flat slab. The horizontal service distribution, construction ease and speed will be made worse if downstand beams are introduced (Figure 4.9).

Figure 4.9 (left) Two-way slab – coffered; (right) performance rating.

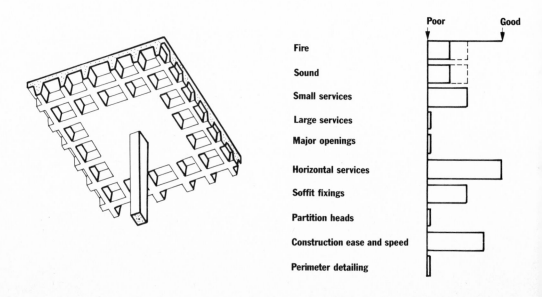

	Poor	Good
Fire		
Sound		
Small services		
Large services		
Major openings		
Horizontal services		
Soffit fixings		
Partition heads		
Construction ease and speed		
Perimeter detailing		

4.5   Performance: Flat Slabs

Solid slab
There are no drop panels or column heads (Figure 4.10).

Figure 4.10   (left) Flat slab – solid; (right) performance rating.

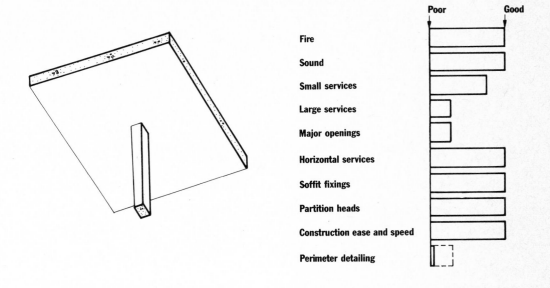

Solid slab with drop panels (Figure 4.11)
If mushroom column heads are used instead of drops, both vertical and horizontal services may be accommodated more easily.

Figure 4.11   (left) Flat slab – solid with drop panels; (right) performance rating.

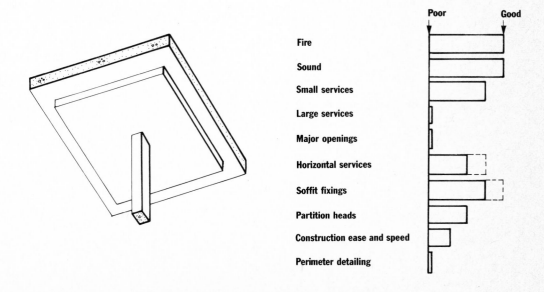

Coffered (waffle) slab
The equivalent of drop panel strengthening is achieved by omitting the coffers around the columns (Figure 4.12).

Figure 4.12   (left) Flat slab – coffered; (right) performance rating.

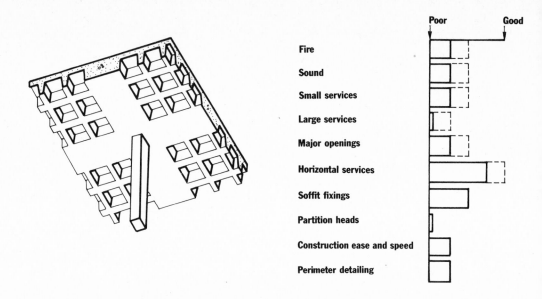

4.6   Preliminary Sizing of Floors

The charts (Figures 4.13–4.15) indicate general arrangements, typical sections and likely ranges of structural floor depths for given spans, covering one-way spanning, two-way spanning and flat slabs. As the different features of the design of each project will influence the choice of suitable floor depths, the plots of depth versus span have been deliberately blurred in order to generalise them. However, they provide an initial assessment of likely slab depths.

For two-way slabs (Figure 4.14) the critical span is the *shorter* of A and B. If one is more than twice the other the slab is effectively one-way spanning. For flat slabs (Figure 4.15) the critical dimension is the *longer* of A and B.

Note that heavily loaded floors and those carrying many partitions will tend to be deeper than similar floors carrying open plan areas, with light superimposed loadings (otherwise partitions crack).

In general 'continuous' slabs, that is those with two or more spans, will be shallower than single-span slabs.

Figure 4.13   One-way slab, typical sections, likely depths.

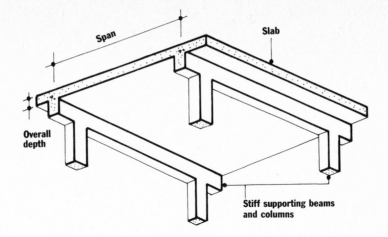

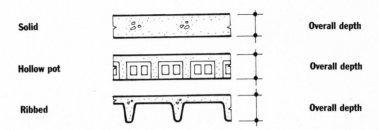

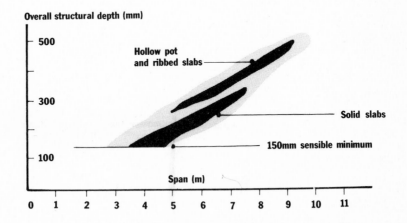

Figure 4.14 Two-way slab, typical sections, likely depths.

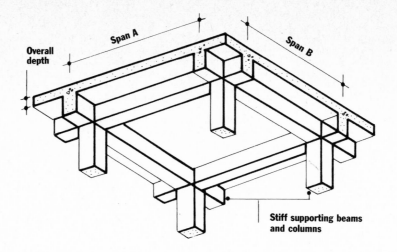

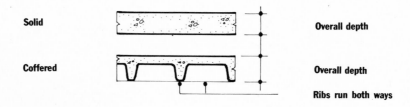

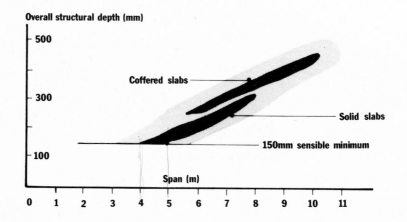

Figure 4.15   Flat slab, typical sections, likely depths.

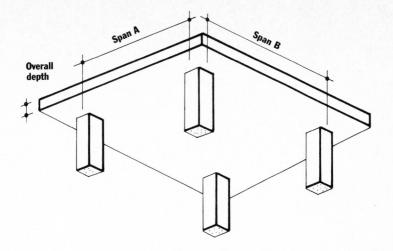

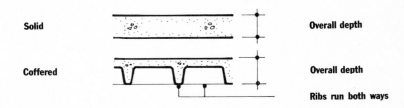

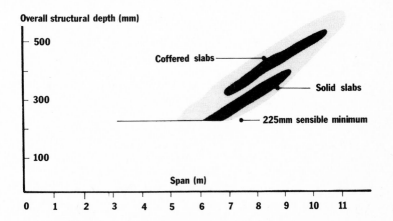

Ribbed Floors

More detailed aspects of ribbed (trough, hollow-pot, waffle or coffered) floors are discussed in this chapter.

Ribbed floors are a lighter form of floor construction than solid slabs of similar strength. This weight saving reduces the loading on the supporting structure and foundations. However, the cost saving in weight must be judged against the extra formwork, fixing and concrete placing costs.

Exposed ribbed floors can give a pleasing visual effect, and the space between can be used to accommodate lighting fittings and other services.

Ribbed floors can be one-way, two-way or flat slabs. For one-way slabs, ribs span in one direction between the beams and walls (Figure 4.5). Where removable formers are used, the resulting floor soffit has a trough form (Figure 4.6). Where floors are two-way spans, ribs can be formed in two directions at right angles to form a waffle or coffered floor (Figure 4.9). Such an arrangement can be designed as a flat slab without intermediate beams, although solid column heads are usually necessary (Figure 4.12).

5.1   Exposed Ribs – Troughs or Waffles

Although moulds for trough and waffle floors can be specially manufactured for a particular configuration of rib depth and spacing, standard moulds for both types of floor are available. They suit the following rib spacing:

o   troughs – 600, 900 and 1200 mm
o   waffles – 600, 800, 900 and 1200 mm.

For each spacing size there is a range of mould depths to suit the varying structural requirements. All moulds are supplied with sloping sides to facilitate withdrawal of the mould after casting.

For ribs up to 900 mm spacing, the soffit width is generally 125 mm, for 1200 mm spacing it is 150 mm. The flange element, (topping) should not be regarded as a screed but cast integrally with the ribs to form a monolithic whole. A minimum flange thickness of 75 mm is recommended (Figure 5.1).

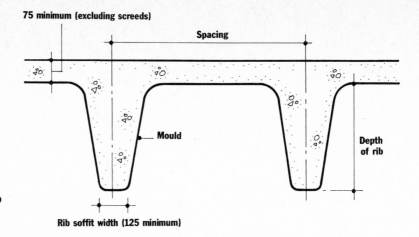

Figure 5.1   General arrangement of exposed rib floor.

5.2   Voided Hollow-pot Floors

In this type of construction the zone between the ribs is formed by a permanent voided element. Traditionally, hollow pots of clay or concrete are used. During the 1960s and early 1970s woodwool formers were used extensively but were found to prevent compaction of the concrete and are not now recommended. Another option, polystyrene formers, has the advantage of light weight, but the formers must be anchored to the soffit during concreting to prevent them floating to the top.

Rib width, depth, reinforcement and aggregate size are all related. BS 8110 states that ribs should be not less than 65 mm wide and their depth not more than four times their width. Although rather narrow ribs can be constructed if done with care, 125 mm is the recommended minimum width (Figures 5.2, 5.3).

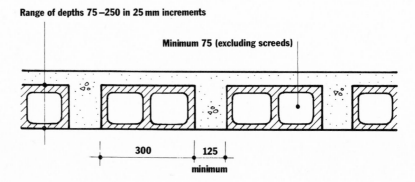

Figure 5.2   Arrangement of normal ribs.

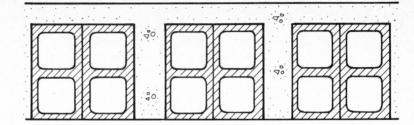

Figure 5.3   Method of forming deep ribs.

Where ribs have been completely covered by permanent formwork there have been problems with honeycombed concrete and displaced reinforcement. This covering should be avoided where possible so that at least the soffits can be seen after the shuttering has been stripped off (Figure 5.4).

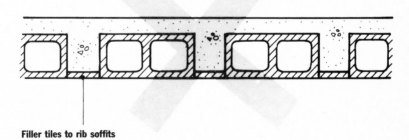

Figure 5.4   Practice not recommended – permanent formwork of tiles obscures view for quality check on rib casting.

Filler tiles to rib soffits

To reduce the overall depth of floor construction, internal beams can be designed as wide ribs with the same depth as other ribs (Figure 5.5). However, it is wise to deepen the edge beams to provide a stiffer base for cladding support around the perimeter of the building.

Figure 5.5   Plan and section of beam created as a wide rib.

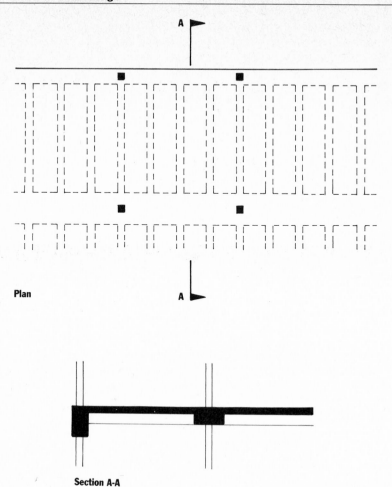

Plan

Section A-A

5.3   Services Penetration

The permissible size and frequency of holes through ribs and flanges must be checked with respect to bending stress, shear in the ribs and deflection. Holes through ribs – usually only relevant to waffle and trough floors – should be circular with a diameter not greater than one-third of the overall depth, and should be placed centrally (Figure 5.6).

In plant rooms and other places where holes exceed the above limits, it may be preferable to use a solid slab.

Figure 5.6 Recommended locations and maximum sizes of holes through flange and ribs.

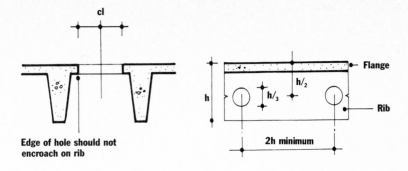

cl

Flange

Rib

h

h/2

h/3

2h minimum

Edge of hole should not encroach on rib

5.4 Partition and Point Loads

Occasionally it is necessary to provide additional transverse ribs to distribute large concentrated loads. However, these extra ribs complicate construction and should be avoided wherever possible by designing the topping to provide this distribution.

Fire resistance
In voided construction, fire resistance can be bridged if partitions are not located at the ribs (Figure 5.7). For exposed ribs (Figure 5.8), where two or more hours' fire resistance is required, regulations may demand an arrangement of secondary reinforcement in the ribs. This is virtually impossible to construct and should be avoided.

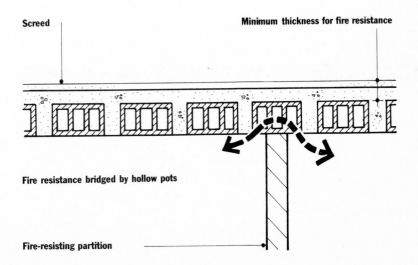

Screed

Minimum thickness for fire resistance

Fire resistance bridged by hollow pots

Fire-resisting partition

Figure 5.7 Path of reduced fire resistance created where partition is located at hollow pot, not at rib.

Figure 5.8   Exposed ribs – fire resistance assessed on minimum flange thickness, including any screed.

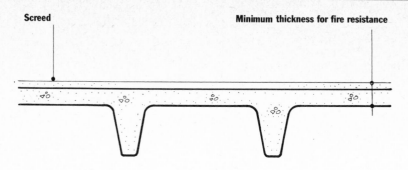

Screed

Minimum thickness for fire resistance

Sound transmission

Locating partitions away from the ribs in voided construction will undermine sound insulation (Figure 5.9). For voided and exposed rib floors (Figure 5.10) sound transmission through floors should be assessed at the thinnest section, not on the basis of an average depth including ribs.

Figure 5.9   Path for flanking sound transmission where partition is at hollow pot, not at rib.

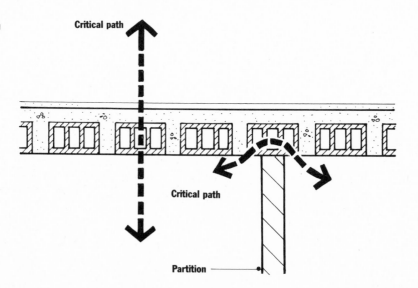

Critical path

Critical path

Partition

Figure 5.10 Exposed ribs: sound attenuation is measured for the thinnest section, that is through the flange.

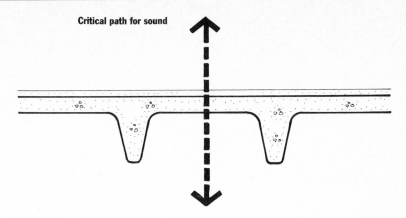

Critical path for sound

5.5 Fixing to Soffits

In voided construction, fixings are not straightforward (Figure 5.11). They include the following:

o   toggle fixing through the pot for light loads such as lighting fittings and cable conduits, but not suspended ceilings

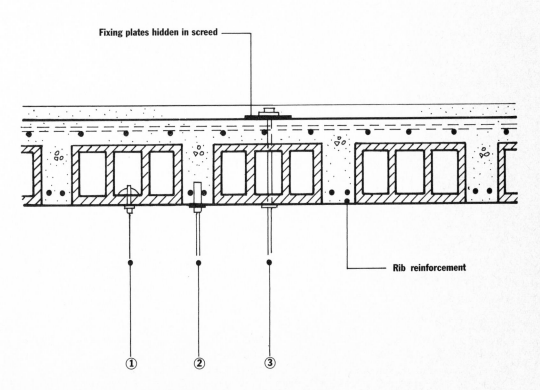

Fixing plates hidden in screed

Rib reinforcement

Figure 5.11 Fixings to soffits in voided construction: 1 – toggle fixing; 2 – expanding anchor bolt or plug and screw; 3 – drop rod.

o   plug and screw or expanding anchor bolt into the centre of the rib – check there is enough space between the reinforcing bars
o   for heavy loads, drop rods can be fitted through the slab, with a fixing plate on top – this is only suitable where screeds cover fixings.

In exposed rib construction, fixing is more affected by the location of the reinforcement (Figure 5.12). Suitable fixings include the following:
o   plug and screw fixing for light loads
o   expanding anchor bolts to the sides of ribs, positioned to miss any reinforcement
o   drop rods from fixing plates on top of the slab, covered by the screed
o   cast-in fixing in the centre of the rib soffit – note that reinforcement will not normally permit drilled-in fixings to rib soffits.

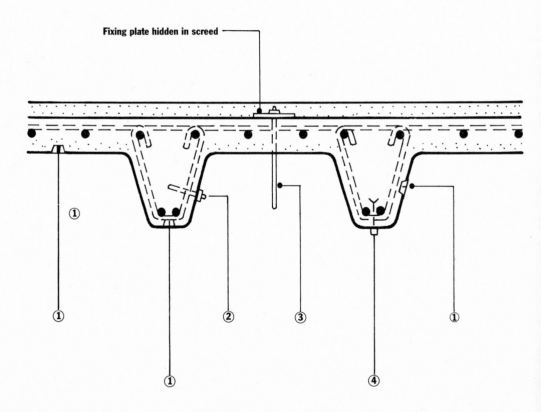

Figure 5.12   Fixings to soffits of exposed ribs:
1 – plug and screw; 2 – expanding anchor bolt;
3 – drop rod; 4 – cast-in fixing.

5.6 Fixing of Partition Heads

Where partitions are stabilised by the slab above, allowance should be made for movement and for avoiding load transfer on to the partition (Figure 5.13). A gap should be provided between the partition and slab. The gap may need to be stopped for fire or acoustic purposes, using a proprietary compressible filler system.

In voided construction a pair of angles can be fixed to the ribs, securing the partition head between them (Figure 5.14).

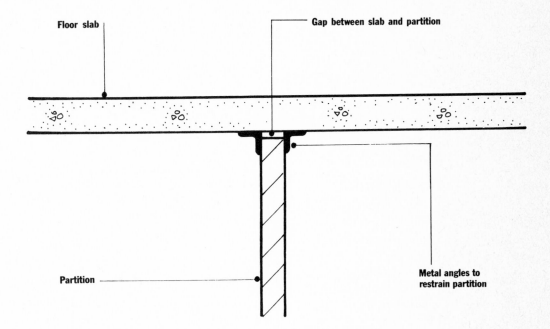

Figure 5.13   Gap engineered between partition and slab, whatever its construction.

Figure 5.14   **Sections at right angles showing metal angles fixed to ribs to secure partition head.**

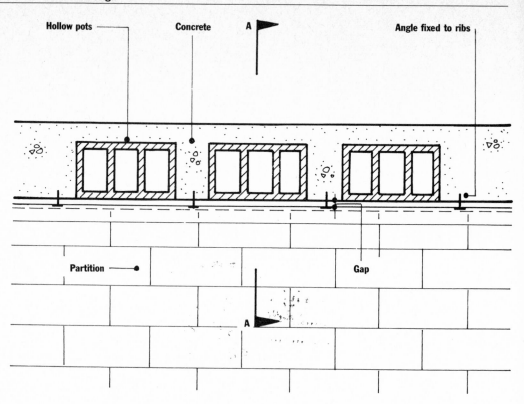

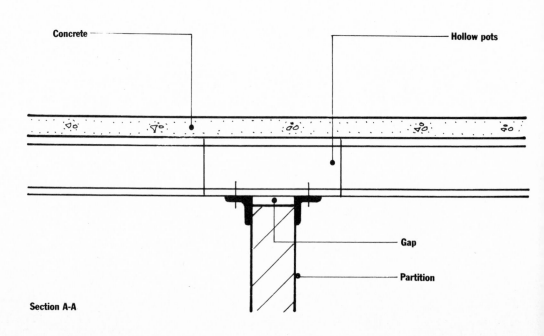

Section A-A

Figure 5.15 Partition parallel to ribs with junction not exposed.

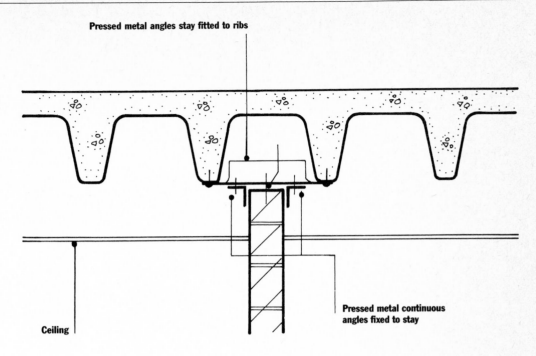

Pressed metal angles stay fitted to ribs

Pressed metal continuous angles fixed to stay

Ceiling

In trough and waffle floors there are several variations, each of which must allow for fire resistance, acoustics and movement where necessary, for example:

o   a partition parallel to the ribs where the suspended ceiling can provide fire and acoustic separation (Figure 5.15)

o   a partition at 90° to the ribs, with a suspended ceiling (Figure 5.16)

o   a partition at 90° to the ribs, where one face of the partition is exposed at its junction with the floor soffit (Figure 5.17) – but this elaborate fitting in of masonry may be badly built. Another option is to cast a cross-rib as part of the floor construction on the line of the partition.

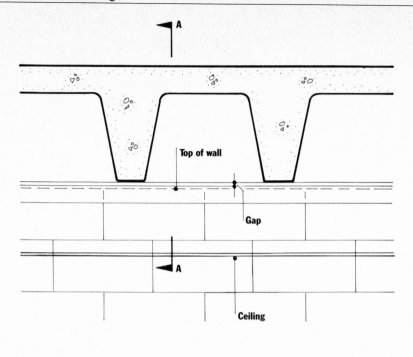

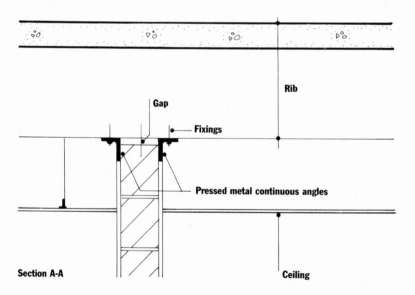

Figure 5.16   Sections at right angles showing fixing of partition head at 90° to ribs, with junction not exposed.

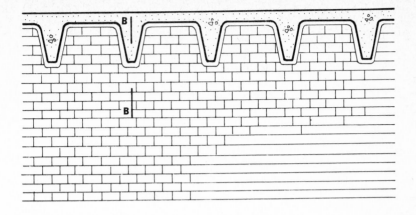

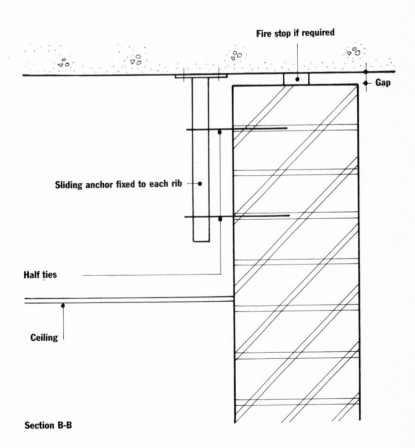

Fire stop if required

Gap

Sliding anchor fixed to each rib

Half ties

Ceiling

Section B-B

Figure 5.17   Elevation and section where partition meets floor soffit and is exposed on one side.

5.7   Points to Note on Site

When laying up void formers in voided construction, gaps between pots are likely to lead to grout loss and should be mortared up (Figure 5.18). This should be included in the specification. It is also necessary to ensure that pots at the ends of ribs have their ends sealed.

For exposed ribs – troughs and waffles – it is important to make allowance for size variations in long runs of moulds (Figure 5.19).

Figure 5.18   Side elevation of a row of hollow pots which do not all abut tightly and so need mortaring up.

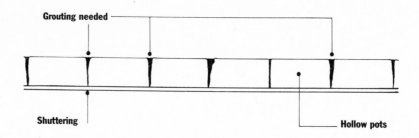

Figure 5.19   The overall length of a long run of moulds should allow for the cumulative effects of size variations. Undersize moulds can be spaced out slightly but a tolerance should be included for oversize moulds.

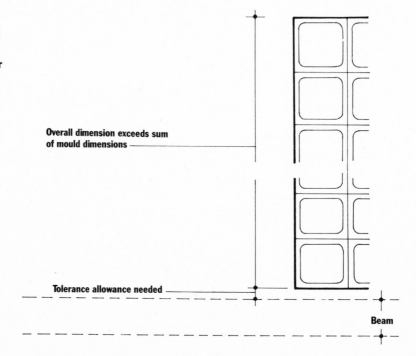

Holes

In Chapter 6, guidance is given on sizes and positions for holes in slabs, beams, columns and walls.

Because holes through reinforced concrete structures pose design, construction and installation problems, the routing of mechanical and electrical services should be planned to minimise conflicts with the structures from the start. Where practicable, vertical distribution of services should be contained in ducts where penetration through the floors can be allowed for in design. Subsequent services modifications can then be contained within the ducts.

6.1 Holes through Beams

As most current contractual arrangements preclude detailed service design at an early stage of building design, it is most important to provide an unobstructed zone for horizontal services distribution. Structures with wide, shallow beams or flat slabs are recommended because these allow horizontal services to be routed below the structure rather than through it (Figures 6.1, 6.2). Above the slab, raised floors can be used (Figure 6.3), or for minor services, proprietary ducts cast into the screed.

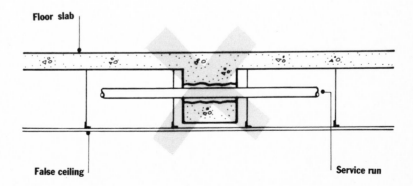

Figure 6.1   Avoid deep beams that require service penetration.

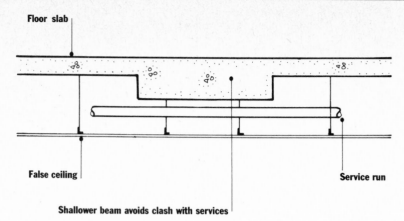

Figure 6.2   Wide but shallow beams overcome the problem shown in Figure 6.1.

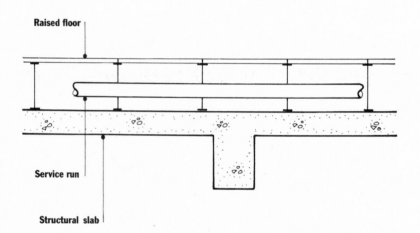

Figure 6.3   The raised floor is a conceptually simple, although not cheap, provision for horizontal services distribution.

Where holes do have to be made in beams, they should be limited to one-third of the beam depth and placed at mid height within the central third of the span (Figure 6.4). Nearer the supports, holes should be smaller and preferably sleeved. Rectangular holes, which can create both structural and construction problems, should be avoided. Ideally holes through beams should be circular and scarce.

Figure 6.4   Preferred arrangement of holes through beams.

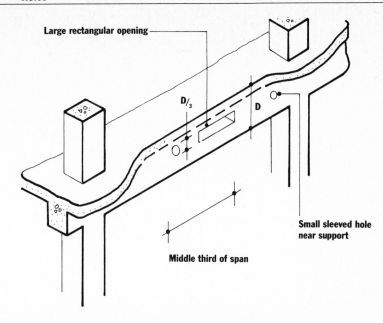

Large rectangular opening

D/3

D

Small sleeved hole near support

Middle third of span

6.2   Holes through Slabs

Major openings for service ducts, lifts, stairs, etc, should be considered at a very early stage as they are intrinsic to the structural design. Usually trimming beams will be needed around openings, or in the case of flat slabs, additional columns (Figures 6.5, 6.6).

Figure 6.5   Slab with extra trimming beams around holes.

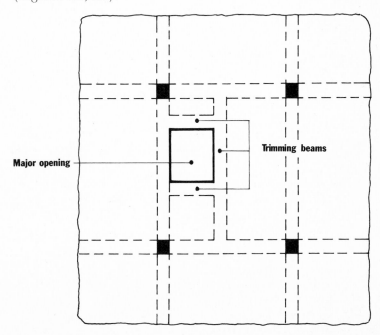

Major opening

Trimming beams

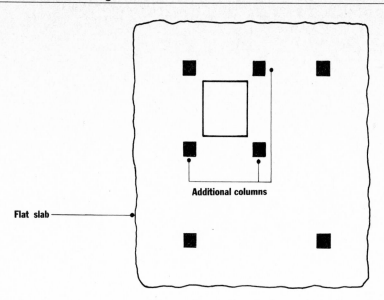

Figure 6.6   In flat slabs (where there are no beams) additional columns are used to provide extra strengthening around major openings.

Medium sized openings (a maximum dimension of up to 1000 mm) can be more easily accommodated as follows:

○   in slabs carried on beams, holes can usually be located anywhere in the slab area (Figure 6.7). There are limits for ribbed floors as discussed in Chapter 5, and holes close to beams may be unacceptable if the beams are designed as T-sections, because the adjacent slab area is used as the top flange of the T (Figure 6.8)

○   in flat slabs, holes should be kept at least one-quarter of the span of the slab away from columns and the slab edge (Figure 6.9)

○   where holes are in groups, they can usually be accommodated if the space left between holes is at least one and a half times the dimension of the larger hole (Figure 6.10). More closely spaced holes should be considered structurally as one opening

○   long slots, usually for electrical services at right angles to the span, should be avoided. A series of circular holes should be used instead (Figure 6.11). It is only possible to cater for slots parallel to the span in one-way slabs (Figure 6.12).

Small holes, usually up to 150 mm, may normally be located in any position, subject to the limitations on grouping given above (see Figure 6.10).

Figure 6.7   Location options for medium sized holes.

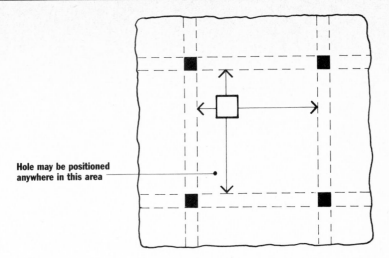

Hole may be positioned
anywhere in this area

Figure 6.8   Limitation on hole location where beam and slab act together as a T-beam.

Hole too near T-beam

T-beam

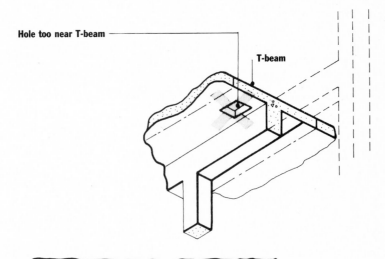

Figure 6.9   In flat slabs, hole location is more restricted.

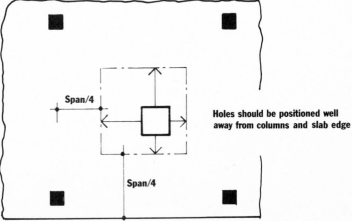

Span/4

Span/4

Holes should be positioned well
away from columns and slab edge

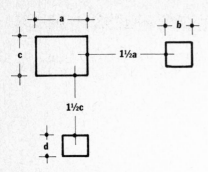

Figure 6.10 Minimum spacing of groups of holes in terms of the size of the larger of an adjacent pair.

Figure 6.11 Avoid slots at right angles to the span: use holes instead.

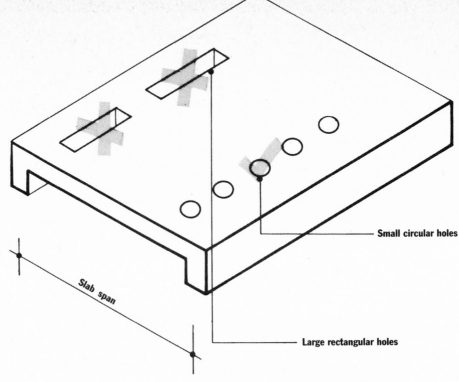

Small circular holes

Slab span

Large rectangular holes

Figure 6.12 Slots parallel to span are only possible in one-way slabs.

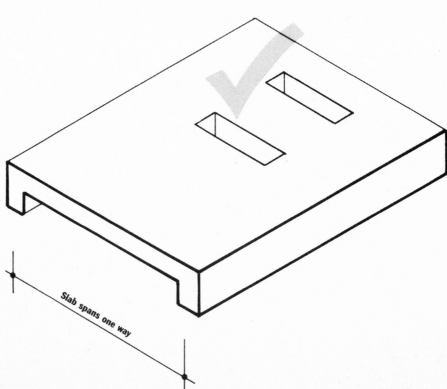

Slab spans one way

6.3 Holes through Columns

Holes through columns should be avoided because even small holes can have a marked effect on structural performance. However, single small circular sleeved holes through the centres of columns may be acceptable (Figure 6.13).

Columns should not be chased for service runs because the chase reduces the structurally effective size of the column (Figure 6.14).

Figure 6.13 Only the smallest of sleeved holes should pass through columns.

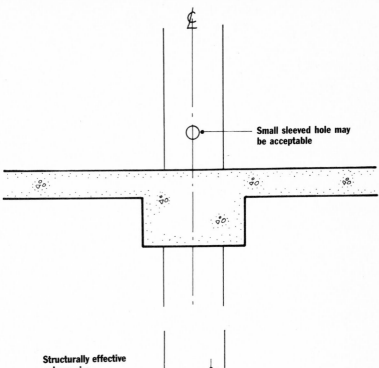

Small sleeved hole may be acceptable

Figure 6.14 Avoid chasing columns.

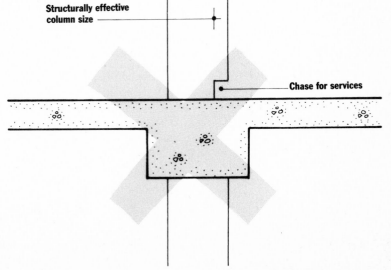

Structurally effective column size

Chase for services

6.4   Holes through Walls

Holes through concrete walls, for instance in plant rooms, should not normally exceed one-fifth of the length or height of the wall (Figure 6.15). Groups of holes should be limited, using the guidance above for slabs carried on beams.

Circular holes are preferred. If rectangular holes are necessary, care is needed on site to ensure that the concrete below the hole is fully compacted (Figure 6.16).

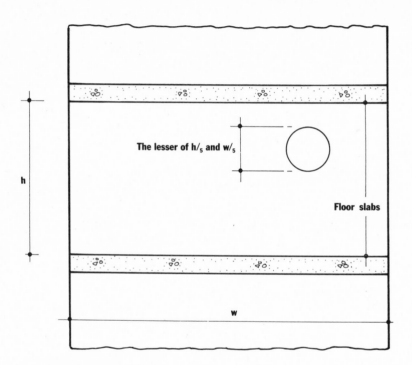

The lesser of $h/_5$ and $w/_5$

h

Floor slabs

w

Figure 6.15   Maximum dimensions for holes through concrete walls.

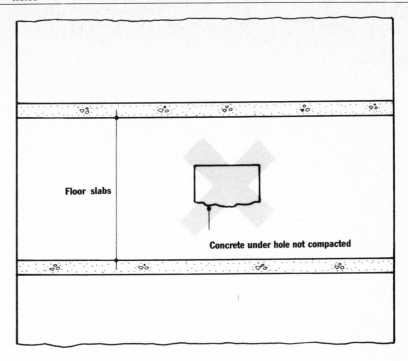

Figure 6.16   Avoid rectangular openings where possible, and where used take care with compaction.

Internal Columns

Chapter 7 looks at the constraints on positions, shapes and sizes for internal columns.

Columns are often the most obtrusive part of a structure and the structural engineer is frequently under pressure to reduce the column size or reposition or reshape a column. He cannot solve these problems by mathematics alone: he has to use his engineering judgement and he may need to embark on wide-ranging negotiations with the design team on construction and economic considerations.

7.1   Column Position

When a column is required to be offset, the use of costly transfer structure is often the only solution (Figure 7.1). Heavy beams have to be introduced, leading to considerable complications in the design, detailing and construction process. The cost implications should be fully investigated. The deeper structure can affect service distribution and reduce headroom.

Figure 7.1   Examples of transfer structure to support off-grid columns.

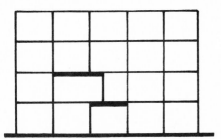
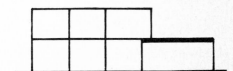

7.2   Column Size

The size of a column may be affected by requirements for durability and fire resistance, which should be considered very early in the design process.

The prime consideration is whether the columns provide lateral stability to the structure as a whole. If shear stability is provided by walls or lift complexes then, using CP 110 definitions, the columns are braced (Figure 7.2a). But if the columns

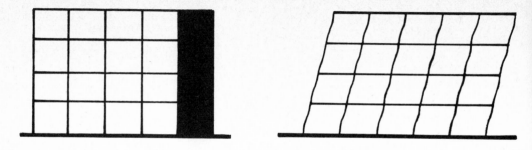

Figure 7.2a   Frame sway-braced by core.   **Figure 7.2b   Unbraced frame.**

have to provide lateral stability they are considered unbraced (Figure 7.2b). Not surprisingly the columns designed as unbraced usually have to be stronger, stiffer and therefore bigger than braced columns.

A major objective when designing columns is to avoid buckling, which would cause a bending failure rather than a pure compression failure. To control buckling, the column size generally needs to be greater for taller structures, which introduces slenderness as an important design factor. Whether columns are considered slender or not depends on their height, the degree of fixity top and bottom and the cross-sectional area. Slender columns have to be designed for increased bending moments, which means a heavy reinforcement and consequently complicated design and difficult construction.

The other factors influencing column size are, of course, the loads to be carried and the way in which the columns are to be loaded – by symmetric or asymmetric structure.

The overall practicability of construction must be considered when deciding column size. One useful method of reducing reinforcement congestion at splices is to use end-bearing square-cut bars, held in concentric contact by welded sleeves or mechanical devices (Figure 7.3). This type of detail is generally more expensive than conventional laps. Another solution for minimum sized columns which have heavily congested reinforcement is to specify a concrete mix of, say, 10 mm maximum sized aggregate rather than the more usual 20 mm maximum size.

Column design is often complex and time-consuming, and there may be many good reasons why a column cannot be reduced below a certain size. From purely practical considerations a freestanding internal column should never be less than 225 × 225 mm, or, where built into a compartment wall, not less than 300 × 190 mm – the latter dimension within the thickness of the wall. These sizes are minima. Generally sizes at least 50 mm greater should be used.

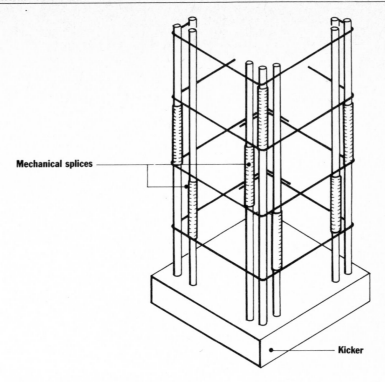

Mechanical splices

Kicker

Figure 7.3   Use of mechanical splices to reduce reinforcement congestion.

7.3   Column Shape

Concrete can be moulded into almost any shape, but there are obvious economic advantages in adopting simple shapes with a reasonable degree of repetition.

Column-beam junctions are usually highly congested with reinforcement (Figure 7.4). If the column shape is the same above and below the junction, the detailing and construction problems are fairly straightforward.

The column's shape and its relationship to other structural elements is also very important. For instance, column reinforcement should continue through the beams without being cranked, and therefore it is desirable to avoid columns and beams with the same widths (Figure 7.5).

If the column is required to offset or to change plan shape between one storey and the next, reinforcement may be impossible to fix and there may be a high risk of poorly compacted concrete in a critical part of the structure. This problem can often be eased by having enough beam depth for the use of lapped splices (Figure 7.6).

Figure 7.4 (this page)   Typical reinforcement congestion at column/beam intersection (based on *Standard Reinforced Concrete Details, Concrete Society, 1973*).

Figure 7.5 (opposite, left)   Beam and column width are the same with the column reinforcement badly arranged, a. Preferred arrangements where the column reinforcement continues straight through the beam, either a column wider than the beam, b, or a beam wider than column, c.

Figure 7.6 (opposite, right)   A deep beam allows a column splice within its depth (based on *Standard Reinforced Concrete Details*).

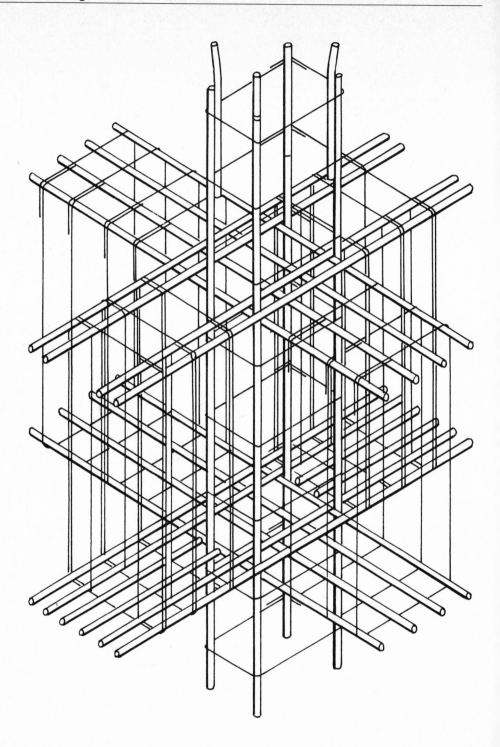

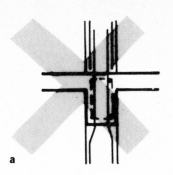

a

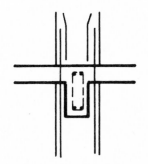

b

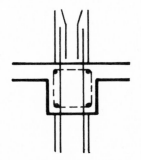

c

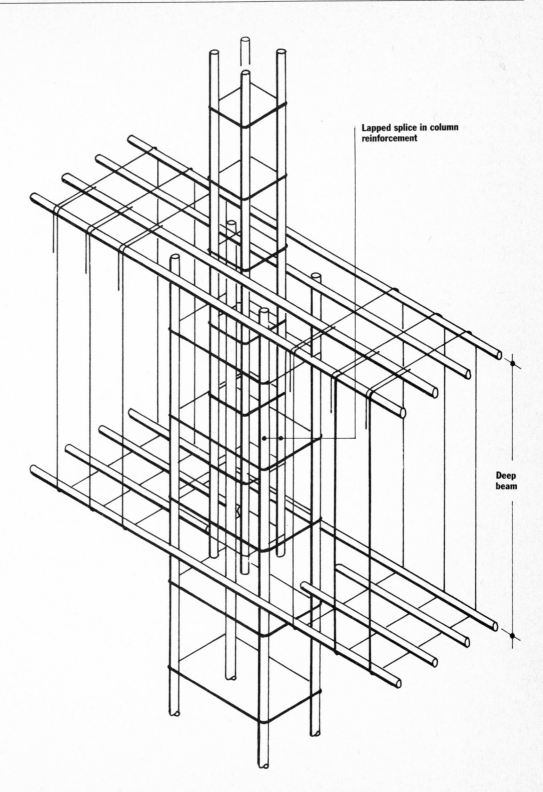

Lapped splice in column reinforcement

Deep beam

17

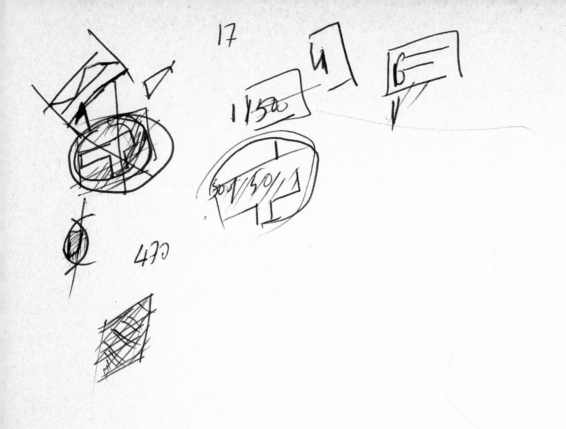

1/500

470

Columns in External Walls

With perimeter columns, the main concerns are junctions with frames, slabs and cladding, and damp-proofing.

Reinforced concrete columns occur in external walls not only as part of the main frame, but also where it is necessary to provide supplementary support to masonry walls. Such columns, or posts, most frequently occur in the external walls where either the size of the masonry panel and/or its boundary conditions (say, windows) are such that the wind forces cannot be contained by the masonry alone.

It is also sometimes necessary to introduce such columns into masonry walls to meet the requirements of A3 of the Building Regulations, which cover the prevention of progressive collapse. In this case these columns are usually required in gable end walls as strong point columns, where it is impracticable to devise any other path of support within the structure.

Reinforced concrete columns in external walls of buildings may cause structural and construction detailing problems which must be considered by the architect and the structural engineer at the earliest possible stage in the design of the building.

Problems either of expressing the slab edge – weathering, cold-bridging, masking with tiles or special bricks, and so on – or of supporting the outer leaf on metal angles are discussed in Chapters 9 and 10.

8.1 Edge Beam – Column – Wall Junctions

The relationship between external columns, edge beams and walls poses several constructional detailing problems, which can be compounded where transverse beams are needed (Figure 8.1). It is necessary to resolve the position of the edge beam in relation to both the column and the external wall, and to decide whether the beam is expressed externally or is contained within the confines of the external wall (see below). Structural aspects must be considered both in relation to shear and the disposition of reinforcement.

Figure 8.1   The intersection of the edge and transverse beam with a column.

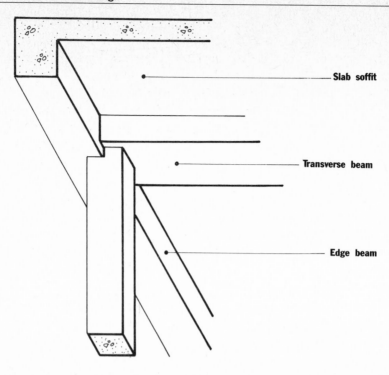

Slab soffit

Transverse beam

Edge beam

8.2   Flat Slab – Column – Wall Junctions

The relationship between external columns, flat slabs (without edge beams) and walls poses not only constructional detailing problems but can also present nearly insoluble structural problems.

Column within slab
Structurally this is the preferred arrangement (Figure 8.2). The dimension

Figure 8.2   Preferred structural arrangement.

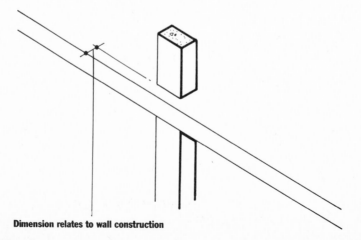

Dimension relates to wall construction

between the external face of the column and the slab edge is geared to external wall construction. A basic decision is whether to express the slab edge on the elevation. Where this is done, measures to overcome cold-bridging may be required.

Column on edge of slab

The reduced slab perimeter around the column could create shear problems (Figure 8.3). This design will require different support systems for inner and outer leaves of masonry cavity construction.

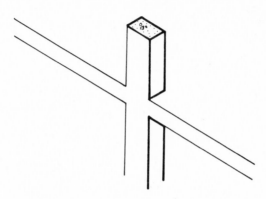

Figure 8.3   A possible structural arrangement.

Column partly outside slab

Structurally this is probably impossible unless floor loading is light and the slab span relatively short or the slab is much deeper than normal (Figure 8.4). There could also be insurmountable problems in supporting masonry cavity construction.

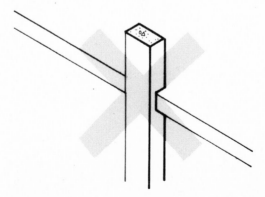

Figure 8.4   An undesirable structural arrangement.

8.3 Wind Posts in Masonry Cavity Cladding

For aesthetic reasons it is unusual for posts to be in the outer skin. Hence wind forces are transferred by ties to the inner skin which acts as a panel in transferring the wind loading to the post. The post acts as a vertical beam to transfer wind loading to horizontal elements of structure – that is, floors or roof (Figure 8.5).

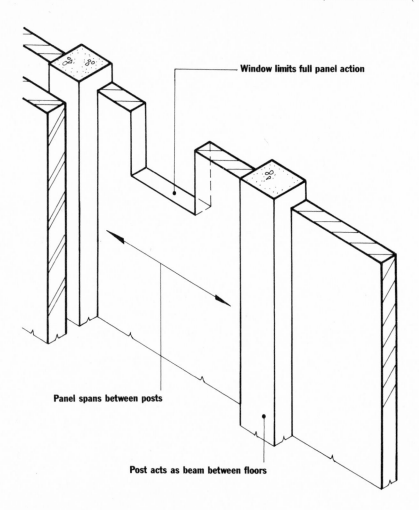

Window limits full panel action

Panel spans between posts

Post acts as beam between floors

Figure 8.5   Wind post gives stiffness to a masonry clad building.

8.4   Strong Point Columns in Loadbearing Masonry

Strong point columns provide an alternative path of support if a section of loadbearing wall should blow out. This is a requirement for buildings over four storeys (Figure 8.6).

Figure 8.6   A strong point column is needed to duplicate the loadbearing capacity of the masonry panel.

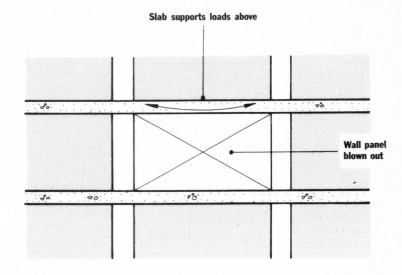

Slab supports loads above

Wall panel blown out

8.5   Columns within Walls

8.5.1   Sizing

For ease of internal planning it is often desirable to size the column so that it fits within the wall construction. This is not normally possible in a traditional wall, because the limitation on column size would significantly reduce its loadbearing capacity.

However, for the supplementary type of column referred to above, where these do not carry vertical loads, it may be practicable to construct such columns without projecting into the internal accommodation. Nevertheless even with these columns it will probably be necessary to project them into the cavity, which causes problems with damp-proofing and cold-bridging. For example:

o   in a traditional 265 mm thick wall the column is much too thin even if extended into the cavity – it may suffice as a supplementary column (Figure 8.7)

o   thicker walls using 140 mm blockwork in the inner skin provide better thermal performance. The column is still likely to be too thin except in a supplementary role, unless extended into the cavity (Figure 8.8).

For economy and convenience in masonry construction, the column dimension and position in the plane of the wall should relate to that of the masonry unit (Figure 8.9).

Figure 8.7   A column fitted within a normal cavity wall is too thin for normal structural purposes.

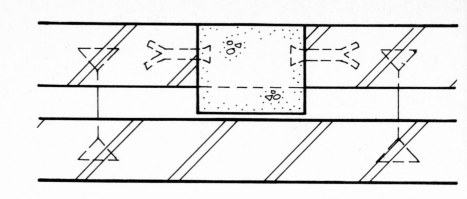

100

60

100

Figure 8.8   Even with a thicker masonry inner leaf than that shown in Figure 8.7, the column thickness still limits its structural use. Wall ties need to be stiff enough to transmit wind loads. Double triangle ties are adequate, but not butterfly ties.

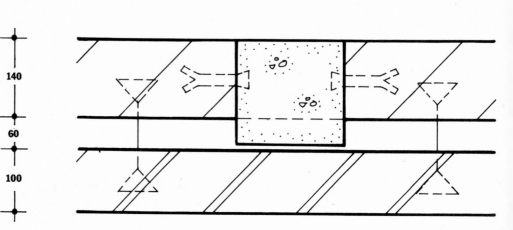

140

60

100

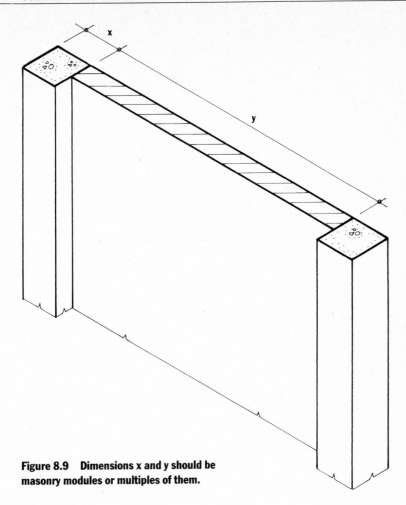

Figure 8.9 Dimensions x and y should be masonry modules or multiples of them.

8.5.2 Robustness

For general robustness and practicability of construction the minimum dimension should not be less than 150 mm, excluding tuck-ins for dpcs and so on (Figure 8.10). When dealing with the less substantial supplementary columns it is important to remember that during construction, when the column is still green and before top restraint is built, the column is vulnerable to accidental forces. A minimum of four bars of reinforcement gives greater robustness (Figure 8.11).

Figure 8.10   The minimum effective column dimension.

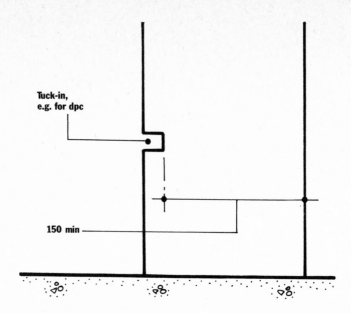

Tuck-in,
e.g. for dpc

150 min

Figure 8.11   Section through a supplementary column. A minimum of four-bar reinforcement provides strength in two planes.

8.5.3   Post-to-Structure Junctions

Where it is necessary to prevent vertical loading of a supplementary column, it is necessary to devise a connection which allows vertical movement but which provides effective horizontal restraint (Figure 8.12).

Figure 8.12   Horizontal restraint without vertical loading of the supplementary column.

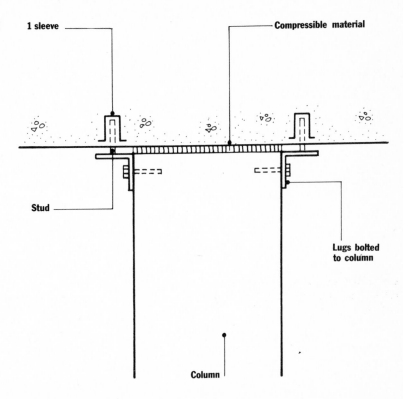

1 sleeve

Compressible material

Stud

Lugs bolted to column

Column

8.6   Columns Encroaching into Rooms

The architect must accept that in most cases columns will project into rooms. Cracking at the junction of different materials is virtually impossible to eliminate.

Flat ties, as shown in Figures 8.13–8.15, are often recommended, but they are difficult to install to ensure adequate lateral restraint. Fishtail ties are easier, although there is a slight risk that the inevitable crack will follow the blockwork bond and not occur in the corner. Corner reinforcement to column plaster is essential. Bed joint reinforcement in the blockwork at critical points such as lintel bearings will help to minimise cracking.

In brickwork cladding there is differential movement as bricks expand and the concrete frame shrinks. Where there are long lengths of wall between columns, movement joints may be necessary, even on the inner skin, although there

**Figure 8.13   A column in blockwork cladding.
A, flat tie; B, cracking follows blockwork bond;
C, corner reinforcement to plaster; and D, fishtail
or double triangle tie.**

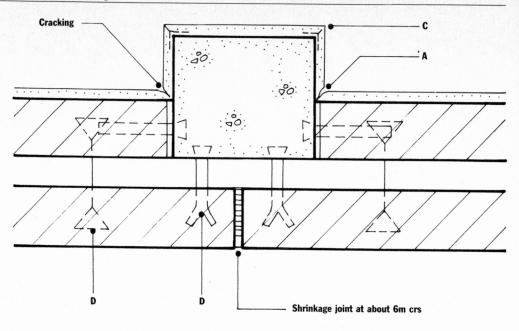

**Figure 8.14   There is a possibility of cracks at
bond joints, rather than at the column-masonry
junction. Key as for Figure 8.13.**

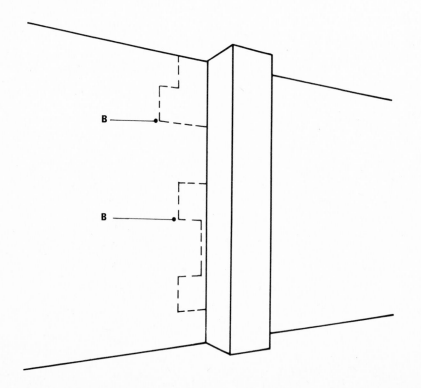

Figure 8.15 A column in brick cladding. Key as for Figure 8.13.

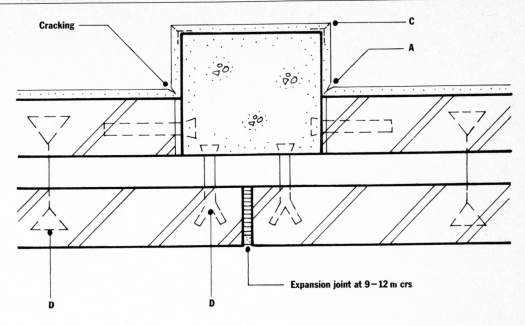

Cracking

C

A

Expansion joint at 9 – 12 m crs

D   D

generally seem to be fewer problems of plaster cracking on a brick backing than on concrete blocks, particularly aerated concrete blocks. This is because the expansion movements of brickwork are generally less than the shrinkage movements of blockwork.

Face brickwork and blockwork is often tied to columns with fishtail ties. In long runs of brick or block cladding differential movement may cause cracking at these ties and more flexible double triangle wire ties might be better.

The shot-firing of pins to fix cramps is not recommended (Figure 8.16).

Figure 8.16 Shot-firing is a poor method of anchoring ties to concrete.

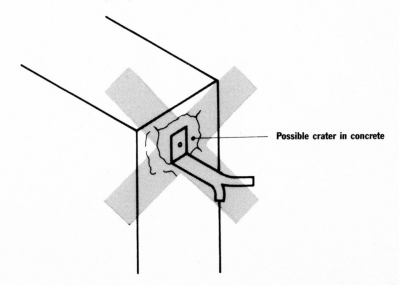

Possible crater in concrete

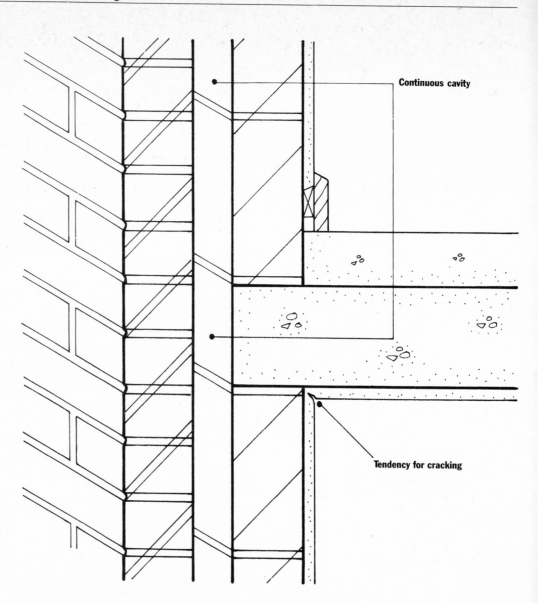

Continuous cavity

Tendency for cracking

Figure 8.17   This is good from a damp-proofing standpoint because it maintains a clear cavity.

8.7   Damp-proofing in External Walls

The complexity of damp-proofing greatly depends upon the relationship of the column and slab in the external wall. If possible, both the column face and slab edge should be kept to the back of the cavity so that the cavity is maintained (Figure 8.17). In this case no damp-proof trays are necessary (for instance, where the masonry cladding must be supported by the structure; see Chapter 9).

8.7.1 Damp-proofing and Insulation

If the slab and/or the column bridges the cavity, great care is necessary to form the damp-proof trays and vertical dpcs (Figures 8.18–8.20). If cavity insulation is specified, careful consideration of details at columns will be necessary if these columns span the cavity to any extent.

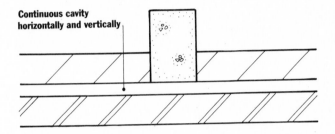

Continuous cavity
horizontally and vertically

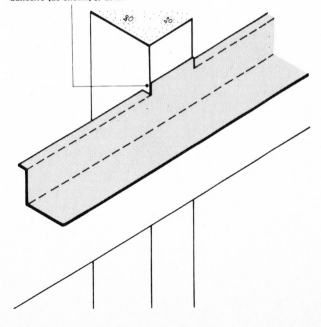

Vertical flap fixed to column by
adhesive (as shown) or at tuck-in

Figures 8.18–8.20   Plans and axonometrics of damp-proofing for typical column-beamed intersections.

Any tuck-in groove reduces effective column size

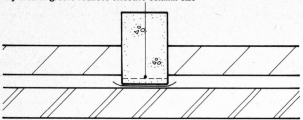

Water check grooves reduce effective column size

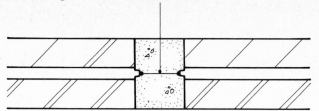

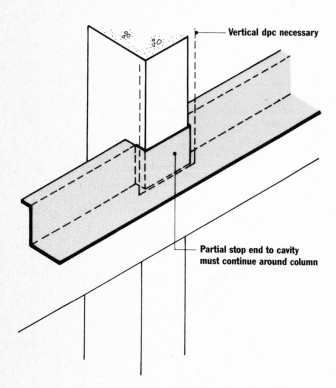

Vertical dpc necessary

Partial stop end to cavity
must continue around column

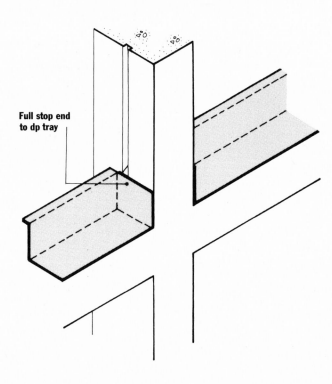

Full stop end
to dp tray

Tuck-ins in the concrete columns reduce their effective size – carefully sticking the dpc to the column may be an alternative. In many instances these constructions form cold bridges which require special consideration.

8.8 Formation of Stop Ends

Stop ends are often omitted at columns where the cavity is bridged, and this omission can lead to disastrous consequences. Preformed stop ends are available in some dpc materials (pitch polymer, Figure 8.21, and polypropylene).

Some dpc materials can be folded to form stop ends on site (Figure 8.22).

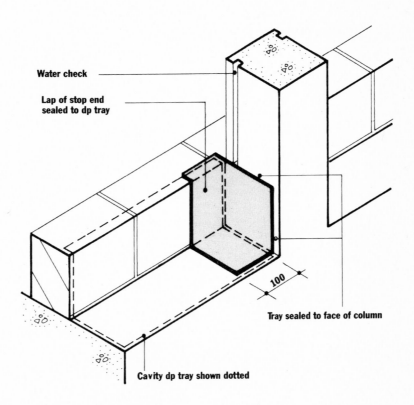

Water check

Lap of stop end
sealed to dp tray

100

Tray sealed to face of column

Cavity dp tray shown dotted

Figure 8.21   Preformed stop end.

Figure 8.22   Stop end formed on site.

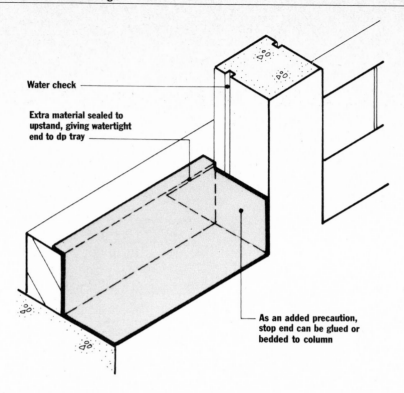

Water check

Extra material sealed to
upstand, giving watertight
end to dp tray

As an added precaution,
stop end can be glued or
bedded to column

8.9   Precast Concrete Cladding

Complex flashings are often required with cladding panels on concrete-framed buildings. Preformed junctions and carefully sealed laps are essential (Figures 8.23, 8.24).

Figures 8.23, 8.24   Typical seals and dpcs for precast cladding.

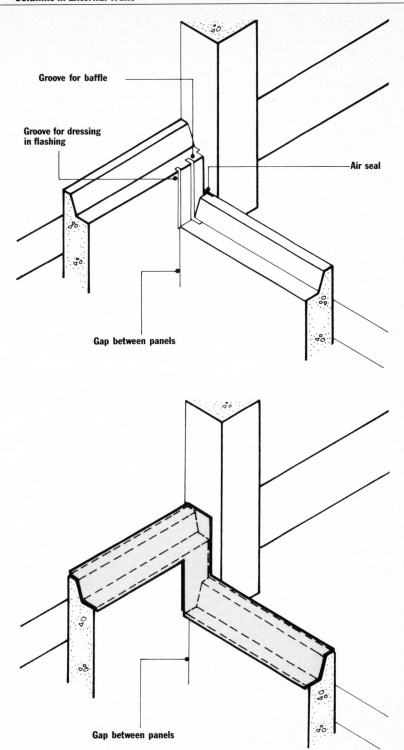

Groove for baffle

Groove for dressing in flashing

Air seal

Gap between panels

Gap between panels

Brick Cladding Support

This chapter goes into more detail on the problems of masonry cladding of reinforced concrete building frames.

There are many failures of stability and integrity of brick cladding fixed to concrete-framed buildings. Differential movement of the two materials has been poorly understood by architects and engineers. The techniques and details developed to accommodate this movement are relatively new and need to be understood by all members of the construction industry. Unfortunately guidance published by various manufacturers is not always aimed at buildability, exclusion of damp and so on.

9.1 How Differential Movement Works

Brick-clad concrete-framed buildings move in two ways that maximise differential movement:

o   the concrete frame shrinks during the process of drying out, is compressed after the application of loads and is subject to long term creep (Figure 9.1)
o   clay brick cladding expands as the newly fired bricks absorb moisture; the amount of expansion depends on the type of clay and firing temperature, and takes place over many years.

Concrete and calcium silicate bricks and blocks shrink after manufacture, so when these materials are used to clad concrete frames there is not so much differential movement. But careful consideration of the cladding details must still accommodate shrinkage, which can be quite large.

9.2 Structural Requirements

Both the old CP 111 and the new BS 5628 codes of practice for the structural use of masonry recommend that external cavity walls are interrupted and supported at intervals to avoid loosening of ties as a result of differential movements. These loosely defined recommendations have recently become a requirement, particu-

Figure 9.1   Differential movements in concrete-framed structure clad in clay brickwork.

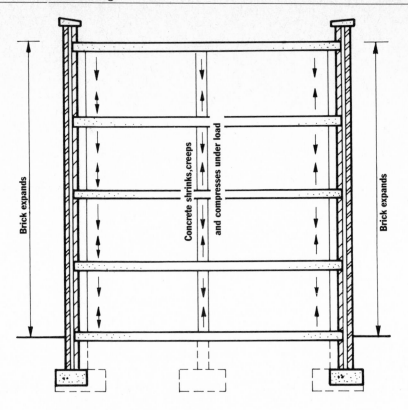

Figure 9.2   Maximum spacing requirement for supporting masonry cavity wall's external leaf.

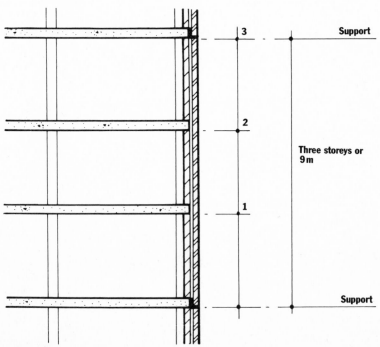

larly for cladding concrete-framed buildings. Figure 9.2 shows the requirements for buildings exceeding four storeys or 12 m, whichever is less. The outer leaf must be supported at intervals of not more than every third storey or every 9 m, whichever is less. Some architects and engineers prefer to support the brickwork at each floor level to reduce the loading on the supporting metal angles in order to minimise the width of movement joints, to standardise the appearance of the building and so on. BS 5628: Part 3, just published and superseding CP 121, gives useful guidance on joint spacing.

9.3　Past Methods of Support

In the 1950s and 1960s the floor slab was often taken through the entire width of the cavity wall and expressed on the elevation. (See Figure 9.3.) This detail is now rarely acceptable because it makes a cold bridge, insulation is difficult to provide and fairfaced concrete has gone out of fashion because of staining and poor workmanship.

Figure 9.3　Support methods for masonry outer leaf: slab revealed on outer face, left; brick slips mask slab edge, right; metal angle (usually stainless steel) now commonly used, over page.

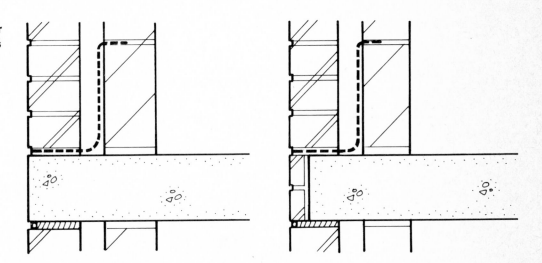

Brick slips have been used to conceal the edge of the slabs (see Figure 9.3) but it is difficult to ensure adhesion of the slips in this case and many failures have been noted. Elaborate mechanical ties have been developed to support and tie back the slips, but the detail is too complex for general application.

Brickwork support on concrete nibs is still used in some cases and is discussed in Chapter 10. The method currently favoured for the support of brickwork on concrete frames is to use metal angles, which are inserted where required in the building. (See Figure 9.3 over page.)

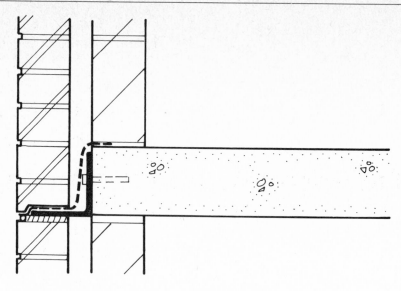

9.4   Use of Stainless Steel Angles

The architect needs advice from a structural engineer on several factors when using stainless steel angles (the numbers below refer to Figure 9.4):

1   the size of the angle depends on the detail – it must support a minimum of two-thirds of the outer leaf thickness. Gusset plates for jointing the angles should be avoided because these conflict with the masonry and the dpc.

2   the gap size for the differential movement should be calculated. It depends on the wall height, brick type specified, frame shrinkage – usually 10–15 mm between joints – and the choice of compressible filler.

3   ties should be located as close as possible to the angle in order to restrain the outer leaf.

4   patent stability ties may be needed to provide lateral support for the wall (see 'Construction details' below on what to do when there is a blockwork inner skin).

5   a movement gap is needed in the inner skin if it is built of brick (see 'Construction details' for blockwork inner skin).

6   drilled fixings should be located to miss the main reinforcement, but they may still hit links (see 'Types of fixing' below).

7   large horseshoe washers may be required to allow for variations in the face of the concrete. It is very important that the angle has a firm backing at its fixings.

8   variations in the position of the angle may be required for architectural reasons such as window heads, necessitating alternative restraint ties.

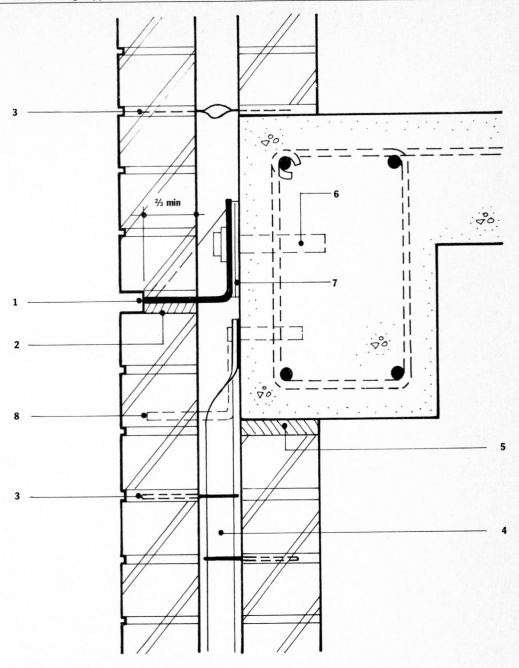

**Figure 9.4   Support of masonry on stainless steel
angle (see text for numbered notes).**

9.5 Types of Fixing

There are two basic alternatives for fixing stainless steel angles: self-drilling bolts (Figure 9.5a) or cast-in toothed anchors (Figure 9.5b). Each type has advantages, and contractors have different views on their merits. The self-drilling bolt does not require careful setting out during the erection of formwork. The toothed anchor slot allows for considerable vertical adjustment later. There is a danger with the bolts that the drill bit will hit a link and have to be offset. There is also a risk that the anchor will become displaced during the casting of the slab or beam.

It is possible to allow for offsets to avoid reinforcement and to take into account building tolerances by providing double holes if self-drilling anchors are being used (Figure 9.5c) or by elongating holes if cast-in anchors are used (Figure 9.5d).

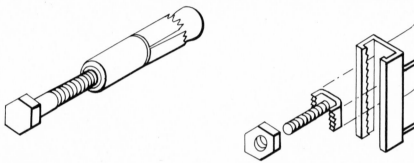

Figure 9.5a   Self-drilling fixing bolt.

Figure 9.5b   Toothed anchor for casting in.

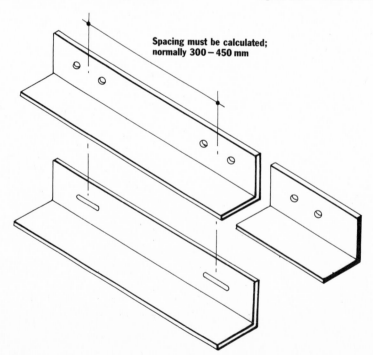

Spacing must be calculated;
normally 300 – 450 mm

Figure 9.5c   Double holes to allow choice of fixing position.

Figure 9.5d   Slots to allow choice of fixing positions.

For brickwork, vertical movement joints are normally 9–12 m apart. Shorter lengths of angle may be used to aid handling. One should carefully consider the levelling of the angle in relation to the brickwork below, otherwise an uneven joint could result.

9.6   Tolerance Problems

Tolerances allowed under current British Standards mean that the edge face of the concrete slab and the brickwork can vary considerably. One may need to define tighter tolerances to obtain a satisfactory job. Figure 9.6, taken from an actual project, shows two extremes. In Figure 9.6a the width of the toe had to be reduced from the design dimension, while in Figure 9.6b the angle had to be mounted on a special continuous packing plate as well as the normal horseshoe washers.

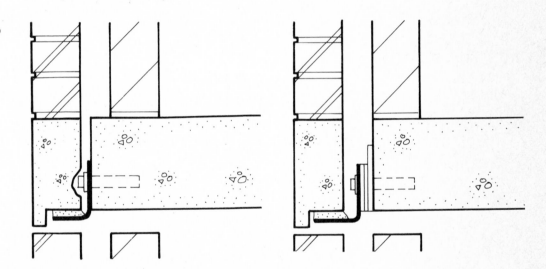

Figure 9.6   Tolerance problems: outer leaf too close to slab, left, and too far away, right.

9.7   Construction Details

Detailing of the support angles is a problem for the architect because it introduces a complex high tech element into more traditional construction. Damp-proofing is made more difficult because the cavity is bridged, often at every floor level. Architects should bear in mind the following points (the numbers given below refer to Figure 9.7):

1   a cavity damp-proof tray is required either with the angle or one course higher to avoid the bolt head. On exposed sites the careful installation, lapping and

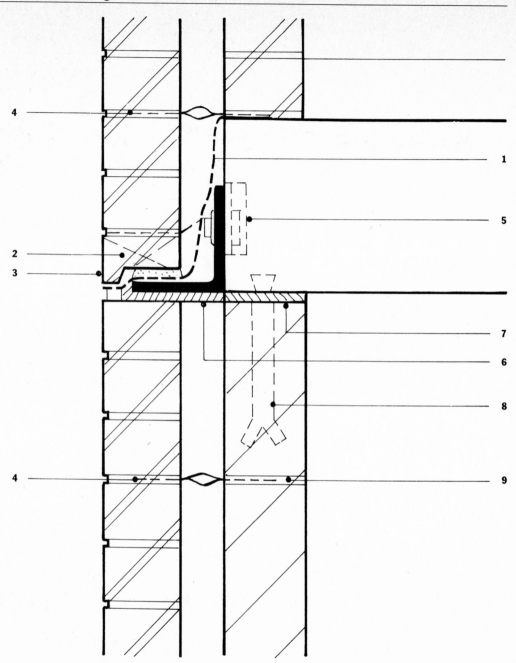

Figure 9.7　Construction detailing at stainless steel angle support (see text for numbered notes).

sealing of the cavity damp-proof tray is essential. If cavity insulation is specified, detailing will be even more complicated. Cleaning cavity trays is difficult; it is probably best to leave mortar droppings unless they are excessive.

2 weepholes should be at 900 mm centres, above the damp-proof tray. Some authorities suggest sticking the dpc to the angle.

3 special nibbed bricks can be produced by many brick manufacturers to minimise the mastic pointed joint. A high performance mastic on half-round foam strip is required to minimise maintenance.

4 cavity ties should be as near the angle as possible.

5 a toothed channel and T-head bolt will allow adjustment for vertical tolerance. Cut off the bolt shank so the damp-proof tray is not vulnerable to damage from the projecting end.

6 the stainless steel angle forms a cold bridge. Extending the compressible filler may overcome the danger of condensation on the underside.

7 blockwork shrinks on drying out, so a movement joint is not normally required in blockwork, unless deflection of the slab is significant.

8 consider a stability tie built into a vertical joint of blockwork, because its fixing does not interfere with the stainless steel angle or bridge the cavity.

9 stability ties may cause awkward cracks at block joints rather than in corners, as normally expected.

9.8 Preformed Drips

The use of a preformed metal drip may be a better long term solution than mastic because less maintenance is required, although sometimes it may not be visually acceptable. Beware of the risk of galvanic action between the metal drip and the angle. A mastic seal may still be required underneath a drip in very exposed locations (Figure 9.8).

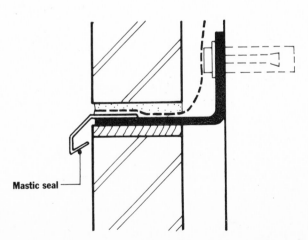

Mastic seal

Figure 9.8   Preformed drip, mastic-sealed if in exposed position.

9.9   Over-complex Details

The reawakening of interest in decorative brickwork must be commended, but many projects are trying to emulate traditional heavy masonry construction in single-skin brickwork. The corbelling illustrated is typical of many such details and is not recommended (Figure 9.9). The fixings are over-complex, there are too many different special bricks and consequently the detail will be very expensive and not very practical for constructing on site. Brick-clad precast concrete panels should be considered as an option.

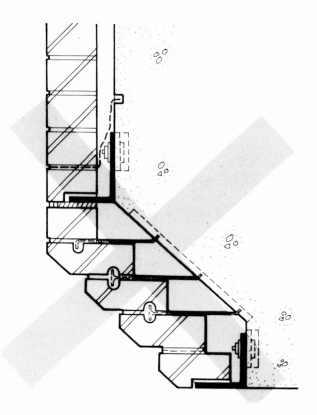

Figure 9.9   Over-complex detail, difficult to build and may be poorly finished.

Concrete Nib Supports

A compromise between appearance and performance is needed when masonry cladding is carried on concrete nibs.

Where cladding is supported on a concrete-framed structure with either minimal or no exposure of the frame, concrete nibs projecting from its floor slabs, edge beams or columns may be used. The size, position and construction details of supporting nibs have an important effect on the structural efficiency of the building as a whole. The following structural points need consideration (the numbers refer to the key numbers in Figures 10.1–10.4):

1 all nib surfaces must be treated as being exposed. Thus 40 mm minimum cover to all reinforcement will normally be required, including all drips and inserts.

2 U-bar nib reinforcement is recommended, preferably in the form of closed links tied to the main links. This forms a rigid cage allowing more accurate casting. A minimum nib depth of 140 mm is recommended to accommodate this arrangement.

3 when U-bars are used, ensure there is adequate depth of anchorage within the beam reinforcement cage.

4 provide restraining cavity ties as close above the support of the cladding panel as possible to ensure adequate stability.

5 it is recommended that the supported cladding should bear over the nib reinforcement by at least 50 mm. Therefore the maximum overhang for a brick cladding panel – nominal thickness 100 mm – should be 10 mm. Allowance should also be made for building tolerances, and strict specification is necessary to achieve a suitable detail.

6 practical considerations of concrete construction including aggregate size, compaction and tolerances in shuttering, and bar bending make the use of thin nibs unsuitable. The recommended nib depth of 140 mm should be treated as a minimum.

7 nibs used with beams that have little or no lateral restraint against twisting must be checked for torsional stability. In this case long spans should be avoided so as to limit beam rotation.

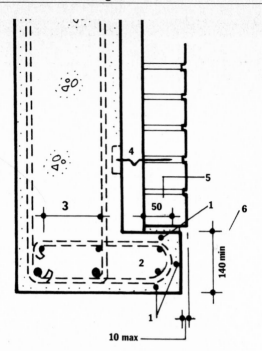

Figure 10.1   Structural aspects of nibs. Dpc omitted for clarity (see text for numbered notes).

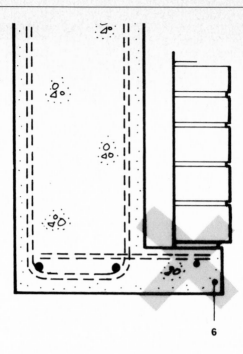

Figure 10.2   Inadequate bearing of cladding over reinforcement. The wall is stable, but the nib is not.

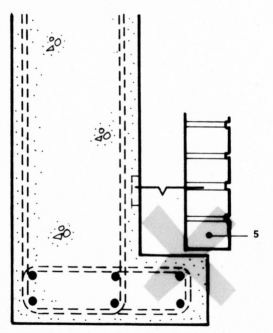

Figure 10.3   The nib is too shallow to accommodate adequate reinforcement.

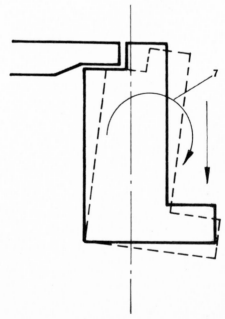

Figure 10.4   Beams with nibs may need special restraint to accommodate torsional loads.

As mentioned above, there are problems with exposed concrete nibs and slab edges, and with masking them with brick slips. Shaped soldier bricks can be used (Figure 10.5). However, this is not recommended because it is difficult to achieve the adequate bearing on the nibs – as noted in number 5 above.

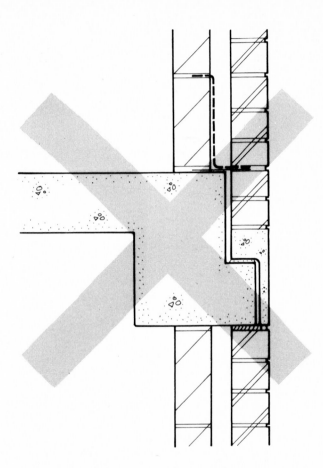

Figure 10.5   Special soldier bricks aim to overcome the problems of slips but tend to reduce the bearing area unacceptably.

In all these cases the effects of cold-bridging must be considered. Insulation may be provided beneath the screed and at ceiling level next to the wall (Figure 10.6). Vapour barriers on the warm side of the insulation may also be required.

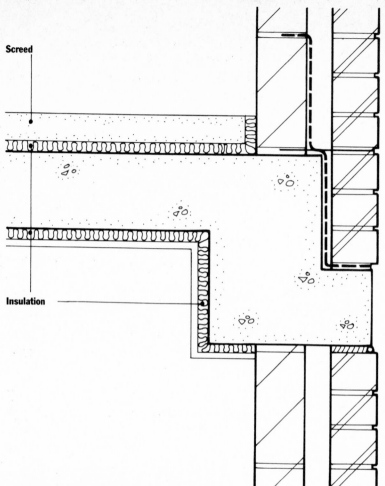

Figure 10.6 Insulation is needed to avoid cold-bridging through concrete.

Where cladding is flush with the nib, use a weather-pointed joint (Figure 10.7a). For a 10 mm overhang use a preformed drip (Figure 10.7b). The choice of damp-proof tray must be compatible with the edge restraint conditions assumed in the cladding design.

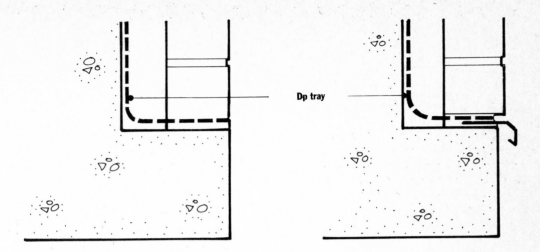

Figure 10.7a   Flush joint, weather struck.

Figure 10.7b   Overhanging masonry with preformed drip.

With deep edge beams, a wall support tie should be fixed to an adjustable channel insert (Figure 10.8). Careful detailing is needed to provide protection for the whole of the beam. Soft joints below beam and nib are essential – usually these are 10–15 mm, but should be calculated. Provide horizontal restraints with vertical adjustment.

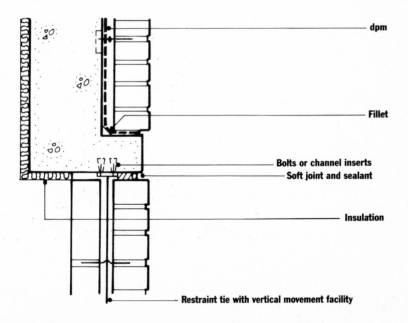

Figure 10.8   Tying masonry above nib and tying in wall below.

10.2   Shuttering

Adequately sized nibs promote good concrete compaction and allow air to escape from the nib section before completing the beam pour. (See Figure 10.9.) Avoid deep narrow beams with thin nib sections. Congestion of reinforcement and the inaccessible position make concrete compaction difficult. (See Figure 10.9.)

Figure 10.9   Size of nib and width of beam determine ease of pouring and compaction. Adequately sized nibs, above, promote good compaction, but narrow beams and thin nibs, below, cause congestion and obstruct compaction.

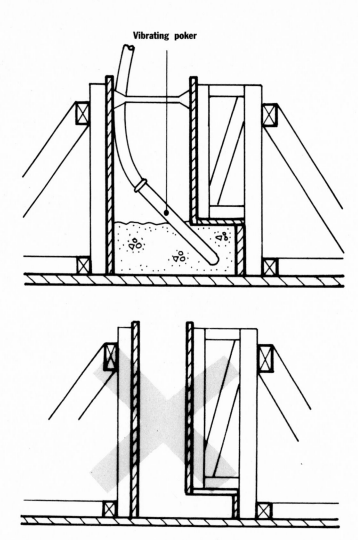

Upstand Kerbs

Another tricky detail occurs when walls are built off roofs or terraces.

Where an external wall is built off a flat concrete roof, an upstand is often formed (Figure 11.1) and problems of construction, waterproofing and insulation have to be solved (the numbers below refer to Figure 11.2):

Figure 11.1   Typical arrangement of upstand on roof.

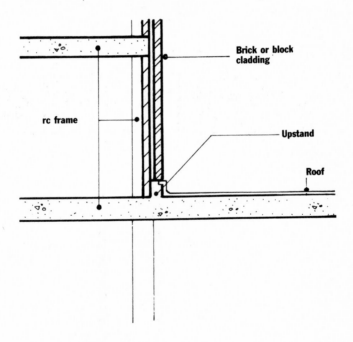

1   the slab must be stiff enough to prevent cracking of the masonry above.

2   sufficient cover to reinforcement must be provided at external surfaces – usually 40 mm.

3   the upstand should be at least 125 mm wide for normal materials and construction.

4   any asphalt tuck should preferably be at the top of the kerb. A tuck lower down increases the necessary kerb width to maintain cover and causes a weak section.

Figure 11.2   Dos and don'ts of structure for kerbs (see text for numbered notes).

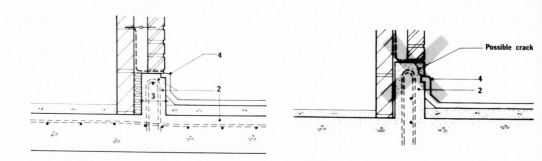

11.1   Construction Details

Detailing follows from the structural points above (numbers refer to Figure 11.3):

1   when detailing the external edge of the dpc a preformed drip should be considered.

2   a 20 × 20 mm rebate is usually the best compromise between asphalting requirements and maintaining kerb width.

3   insulation is needed for cold-bridging behind the kerb and under the screed; a vapour barrier may also be required.

4   the Building Regulations require at least a 150 mm high upstand to roofing.

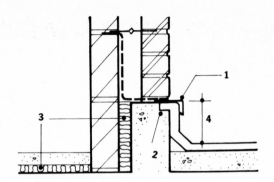

Figure 11.3   Detailing aspects of kerbs (see text for numbered notes).

11.2   Junctions with Columns

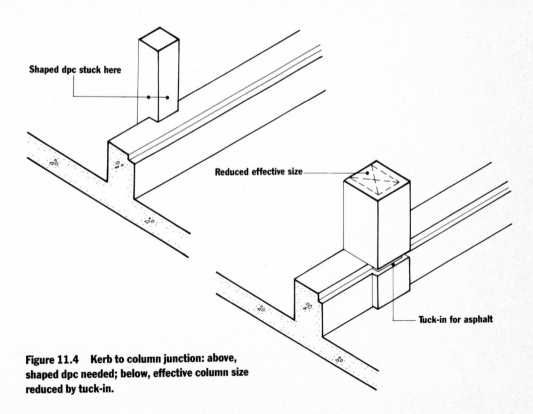

Shaped dpc stuck here

Reduced effective size

Tuck-in for asphalt

Figure 11.4   Kerb to column junction: above, shaped dpc needed; below, effective column size reduced by tuck-in.

Specially formed dpc cloaking will need to be stuck to faces of columns (where the column is behind the outer skin). (See Figure 11.4.) Where the column projects with an asphalt tuck-in taken around the face, the effective structural size of the column is reduced. (See Figure 11.4.)

11.3   Shuttering

Shuttering for kerbs should be kept simple, with a minimum 100 mm opening for placing concrete (Figure 11.5a). Avoid complex rebates, which reduce openings and make compaction difficult (Figure 11.5b).

Figure 11.5   Dos and don'ts to allow convenient casting of kerbs.

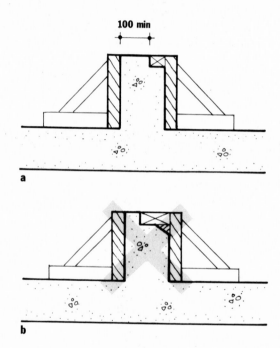

Overhangs

The final chapter looks at the converse of Chapter 11, the overhang.

Forming the overhang in a reinforced concrete building exposes the soffit of the slab and poses problems of weathering, durability, appearance and insulation (Figure 12.1). The solutions to these problems conflict with each other, so the final outcome is a compromise. The following are the key structural points (the numbers refer to Figure 12.2):

1   the slab must be stiff enough to prevent cracking of the very stiff masonry cladding it supports.

2   sufficient cover to reinforcement – usually 40 mm – must be provided at exposed surfaces, inclusive of drips.

3   edge profiles must allow reinforcement to extend under the outer skin of wall.

Figure 12.1   Arrangement of overhang.

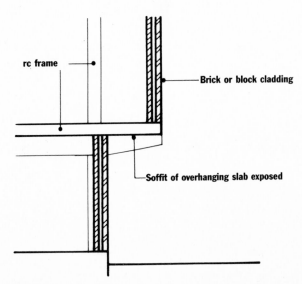

4   opinions differ on the optimum profile for drips. A rounded shape is less vulnerable to spalling arrises. Chamfered arrises are also worth considering.

5   a soft joint and restraint ties are needed at the head of the wall below. A soft joint may not be necessary with a blockwork inner skin – check relative movements.

Figure 12.2   Dos and don'ts for structural aspects of overhangs (see text for numbered notes).

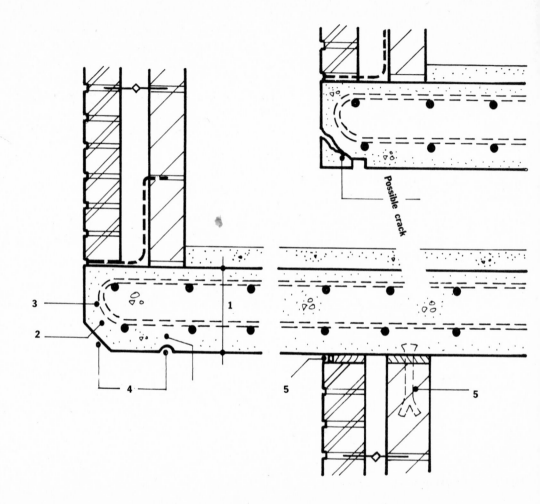

12.1   Construction Details

Detailing follows from the structural aspects above (the numbers refer to Figure 12.3):

1   when detailing the exposed edge of the dpc a preformed drip may be considered in exposed places.

2   the height of exposed concrete slab edge may be reduced by chamfering;

but good lines are very difficult to achieve on site with chamfered arrises – chamfers are likely to stain.

3   a drip must be present to prevent water running back along the soffit. This needs early consideration because of implications for cover and therefore slab thickness.

4   insulation for cold-bridging under the screed will require an extra thickness of screed to accommodate it. There is also a risk of cracking where insulation stops – consider light mesh reinforcement at the transition point.

5   insulating the soffit has problems. It is difficult to carry insulation over the walls below – a durable surface is required. Lightweight insulation tends to float upwards during casting of the slab, and insulation causes a change in reinforcement positions to maintain cover.

6   vapour barriers may be needed to prevent condensation.

Figure 12.3   Constructional detailing aspects of overhangs (see text for numbered notes).

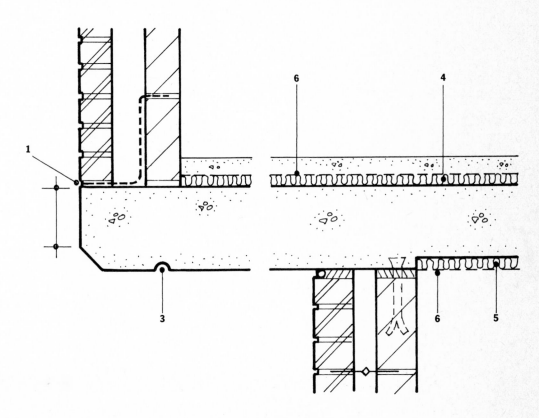

12.2   Shuttering

A flat soffit shutter simplifies site work greatly. Avoid downstand profiles that complicate reinforcement, shuttering and concrete placement (Figure 12.4).

Figure 12.4   Dos and don'ts of shuttering overhangs.

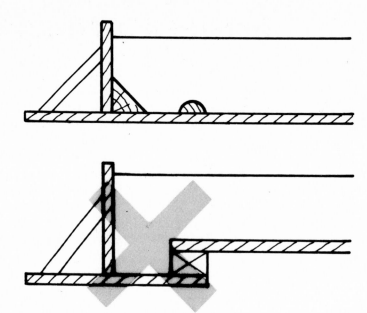

Concrete Buildings – Examples of successful collaboration between architect and engineer

The structures illustrated in the following pages represent a small selection from the vast stock of excellent concrete buildings in which architects and engineers have worked closely together. Successful completion of these important projects is a tribute both to the conceptual thinking that lay behind the designs and to the way this has been followed through into the working details.

Success in the final result requires collaboration between all members of the construction team. Decisions on details, taken jointly between the architect and the engineer have a major effect on all subsequent operations and therefore play a key role. These decisions are frequently bound up with the design and cannot be separated from it or left to a later stage. This fact is highlighted in these examples.

Crown Reach, Grosvenor Road, London SW1

Sixty luxury flats and houses overlooking the river Thames. The building has a complex in situ concrete frame up to nine storeys high and is clad with polished granite, ceramic tiles and brickwork. A very high standard of accuracy and finish was required and achieved.

Architects: Nicholas Lacey with Maguire and Murray
Engineers: Alan Baxter & Associates

Highpoint, Highgate, London

A landmark of the 'thirties, this fine building is shown here photographed in the autumn of 1984. The walls and the frame are of in situ reinforced concrete.

Architects: Lubetkin and Tecton
Engineers: Ove Arup & Partners

Auditorium Ramp, Harrogate Centre

The spiral pedestrian ramp is of in situ
reinforced white concrete and provides a major
feature of the 2,000 seat auditorium and
exhibition/conference centre. The ramp, visible
to the passing public through suspended glass
walls, springs from the main entrance foyer and
rises on an average gradient of 1:13 to the
upper foyer and then onto a viewing platform
before entering directly into the auditorium
bowl.

Architects: Morgan Bentley Ferguson Cale
Engineers: Robert T. Horne and Partners

Headquarters offices for Johnson & Johnson Ltd, Slough, Bucks

The building is arranged in a group of linked octagonal modules on a reinforced concrete frame. To give maximum flexibility of floor usage within the modules the outer structure has been placed on the perimeter of the modules on a square column grid.

Architects: Salmon Speed Architects
Engineers: R. Travers Morgan and Partners

Climatic Research Unit, University of East Anglia, Norwich

The building contains the university's meteorological research facilities and forms the symbolic 'gatehouse' to the newest developments on the campus – the Education Department and the Sainsbury Centre. It comprises four circular floors and is in situ concrete framed with glazed blockwork cladding supported on stainless steel angles.

Architects: Rick Mather Architects
Engineers: Alan Baxter & Associates

Photograph: Patrick Shanahan

Five storey industrial unit for V & E Friedland, Stockport, Cheshire, manufacturer of domestic and industrial sound signalling equipment.

All services and main circulation areas are carried in the in situ reinforced concrete galleried frame, leaving a single volume internal space with provision for a demountable mezzanine floor. The picture shows concrete blockwork outer walls in harmony with the in situ concrete frame.

Architects: Edmund Shipway & Partners
Engineers: W. G. Curtin & Partners

Mobil Court, Strand, London WC2

European headquarters for the oil multi-national company Mobil and the centre for North Sea operations. The building has an in situ reinforced concrete frame and non-loadbearing external cladding of traditional Portland stonework, brick or precast concrete depending on the elevation. Cladding is supported at each floor level by stainless steel angles bolted to the perimeter reinforced concrete edge beams.

Architects: G M W Partnership
Engineers: Clarke Nicholls and Marcel

Cracking up?
Foundation Design

Allan Hodgkinson

Cracking of buildings, because of subsidence
and other types of soil movement, is a major and
common problem. To avoid it requires an adequate
understanding of both soil behaviour and structural
design. This manual offers simple guidance in the form
of practical hints, design data and detailed drawings,
making it an essential aid for architects, builders and
engineers and for students in these disciplines.

1986   128pp   125 illus   234 × 234 mm   cloth
ISBN 0 85139 837 5   **£14.95**

Prices are subject to change without notice

Contents

The Architectural Press Ltd
9 Queen Anne's Gate
London SW1H 9BY

ARCHITECTURAL PRESS